Alliant International University
Los Angeles Campus Library
1000 South Fremont Ave., Unit 5
Alhambra, CA 91803

International Perspectives on Aging

Volume 8

Series Editors

Jason L. Powell
Sheying Chen

For further volumes:
http://www.springer.com/series/8818

Sue Thompson

Reciprocity and Dependency in Old Age

Indian and UK Perspectives

Sue Thompson
Wrexham
Wales
UK

ISBN 978-1-4614-6686-4   ISBN 978-1-4614-6687-1 (eBook)
DOI 10.1007/978-1-4614-6687-1
Springer New York Heidelberg Dordrecht London

Library of Congress Control Number: 2013933573

© Springer Science+Business Media New York 2013
This work is subject to copyright. All rights are reserved by the Publisher, whether the whole or part of the material is concerned, specifically the rights of translation, reprinting, reuse of illustrations, recitation, broadcasting, reproduction on microfilms or in any other physical way, and transmission or information storage and retrieval, electronic adaptation, computer software, or by similar or dissimilar methodology now known or hereafter developed. Exempted from this legal reservation are brief excerpts in connection with reviews or scholarly analysis or material supplied specifically for the purpose of being entered and executed on a computer system, for exclusive use by the purchaser of the work. Duplication of this publication or parts thereof is permitted only under the provisions of the Copyright Law of the Publisher's location, in its current version, and permission for use must always be obtained from Springer. Permissions for use may be obtained through RightsLink at the Copyright Clearance Center. Violations are liable to prosecution under the respective Copyright Law.
The use of general descriptive names, registered names, trademarks, service marks, etc. in this publication does not imply, even in the absence of a specific statement, that such names are exempt from the relevant protective laws and regulations and therefore free for general use.
While the advice and information in this book are believed to be true and accurate at the date of publication, neither the authors nor the editors nor the publisher can accept any legal responsibility for any errors or omissions that may be made. The publisher makes no warranty, express or implied, with respect to the material contained herein.

Printed on acid-free paper

Springer is part of Springer Science+Business Media (www.springer.com)

I dedicate this book to the memory of Ben-Joshua Jaffee, whose insight and generosity of spirit many years ago planted the seed that has grown into an enduring interest in the concepts of reciprocity and dependency

Preface

This book is an account of a research study I undertook at the University of Liverpool, England, between 2008 and 2012, and for which I was awarded a doctorate. Consistent with the university's expectation of reflexivity, I take the reader on a journey—a journey which began with an observation that it is not uncommon for very dependent older people to be conceptualised as one-dimensional—as takers but not also as givers. Along the way, I heard accounts of personal experience that led me to believe that my hunch may have been correct, although for different reasons in different social contexts, and which gave me food for thought about why this should be so and what can be done about it.

Of the many findings, the most striking and pertinent from my point of view was the extent to which the older people in this study had experienced care, however well-intentioned, as often neglectful of the future aspect of their lives—that of aspiration to continue being the person they consider themselves to be now, or to change because of growth rather than decline. It is my hope that, after reading this book, you will be as convinced by their accounts as I am that eldercare theorists, policy-makers, providers and practitioners have much to gain from taking on board the meaning making of those in receipt of care support, as well as conceptualising care from the perspective of their own meaning making.

The book comprises nine chapters, presented in three parts:

In Chap. 1 I discuss how and why the study came into being, I hope it gives a flavour of the passion I have for promoting change through the medium of critically reflective practice, and my aspiration that the findings from this research will play at least some part in that change process.

In Chap. 2 I contextualise the study in a broad-ranging literature base relating to developments in the theorising of ageing, the social significance of old age, policy relating to eldercare in both the UK and India and the main tenets of the phenomenological lens through which I seek to better understand the experience of reciprocity in eldercare.

Building on this, in Chap. 3 I locate my research in the context of literature that relates more specifically to reciprocity itself, and to eldercare within that broader context.

Chapter 4 is the first of Part II, where I expand on the research focus, design, methods and analytical tool, and engage reflexively on issues such as bias, power and research ethics.

In Chap. 5 I present the findings and, given that they emerge from narratives constructed by the participants, I provide as many excerpts from those narratives as is practicable, in order to represent their perspectives as accurately as I can.

In Part III I focus on analysis of the findings, and their implications at a number of levels. I begin, in Chap. 6, by discussing the significance of the findings for the spiritual well-being of older people in receipt of formal care. Structured in line with the four research questions, this discussion highlights most particularly the significance of recognising that dependent older people have a future dimension to their lives, and aspirations associated with it.

In Chap. 7 the focus moves to the significance of the findings for the theorising of ageing, where I locate them in the context of the need to focus on the interaction between agency *and* culture *and* structure in a way which goes beyond:

(a) the atomism of purely psychologically oriented theories;
(b) marxist-influenced structuralist theories which focus on structural issues to the relative neglect of agency; and
(c) post-structuralist and postmodernist approaches that have an emphasis on the cultural/discursive to the relative neglect of agency.

I do this with reference to a further analytical framework—PCS analysis, which I discuss briefly in Chap. 1, and develop in more detail in Chap. 2. As with the four-part phenomenologically grounded analytical framework presented in Chap. 1 (Fig. 1), its bases of dynamism and dialectical interplay also enable a future dimension in older people's lives to be placed on the agenda as a challenge to theorising that has largely neglected it because of an emphasis on decline rather than growth.

In Chap. 8 I build on the discussion in the previous two chapters to draw out the implications for eldercare practice and propose the study's analytical frameworks as learning tools in that context.

In Chap. 9, by way of concluding, I reiterate the significance of the findings and reflect on:

- *why* and *how* they are significant;
- the contribution I have made to extending the knowledge base of reciprocity in eldercare;
- what the findings have not addressed and how this might therefore constitute grounds for further research by myself or others; and
- my own journey through this project, and the part that the participants played in that journey.

I have hopes that the findings, and their significance, will be disseminated beyond the world of academe. The potential to influence eldercare practice, if only in some small way, or as a foundation for future research and development, is what motivates me to research and write. I therefore consider that the value of this research lies in the potential for the two conceptual frameworks presented within it to ultimately operate as aids to understanding that will foster the integration of theory and practice. To some extent, the structuring of Part III may appear, on the face of it, to run counter to my argument for critically reflective practitioners *not* to separate theory and practice into discrete areas (Thompson and Thompson 2008) although I would argue that the two aspects are indeed integrated throughout this work.

I acknowledge that several of the key terms I use in this study are contested and I wish to avoid being led into debates not pertinent to my particular focus because of differences of interpretation (Sunderland 2004). Therefore, prior to the main body of the book, I provide a glossary of key concepts to clarify how I have interpreted and used them here. Placing this glossary at the beginning of the book is perhaps unusual, but I feel it is important that the terms I use are understood in the sense that I have intended them to be. I therefore urge you to read the Key Concepts section before moving on to the introductory chapter.

Key Concepts

Alienation: This refers to a sense of 'apartness', which Crossley (2005) describes as: 'a separation or estrangement of human beings either from each other, from their own life or self, or from society' (p.3). This would seem to draw on the following definition by Fromm (1955), who also refers to 'estrangement' from oneself: 'By alienation is meant a mode of experience in which the person experiences himself [sic] as an alien. He has become, one might say, estranged from himself' (p. 110). I use it, then, with reference to the experiencing of an internal contradiction between the person a dependent older person feels him or herself to be (that is, capable of 'usefulness' through reciprocity) and the person he or she feels that others are perceiving him or her as being (just another older person to be looked after).

Dependency: I am also aware of the contested nature of the term dependency, and would agree with those who argue that it is socially constructed (Wenger 1986; Fine and Glendenning 2005; Grenier 2007; Townsend 2007; Carney 2010). Given that we are all dependent on others in many ways and points in our lives, and that I propose that being able to give as well as receive is not age specific, I do not wish to support the ageist premise that dependency in old age is normative. I am not totally comfortable with using the term dependent, therefore, to describe those elders whose stories I listened to because, while they were dependent on others in some senses, they had much to give in others. However, I considered it to be a necessary evil because my research focus is on those elders externally defined as dependent on formal support as, in the UK at least, being considered to be very dependent has become a prerequisite for attracting formal care support because of increasingly stringent eligibility criteria (Grenier 2007; Lymbery 2010).

Elders: I am aware that old age is a highly contested, socially constructed and fluid concept, open to objective and subjective interpretation (Walker 2005; Townsend 2007; Bytheway 2011). In the context of a current wave of research which highlights positive aspects of ageing (Gilleard and Higgs 2005; Hill 2005; Haber 2009) I have been interested to explore whether those who have become dependent on formal care support feel excluded from such conceptualising of old age as being characterised less by deficit and more by engagement and contribution. While significant dependency is often associated in the research field with those aged 85 and over, often described as the '4th age' (Bond and Cabrero 2007), I have chosen a broader age range from which to draw the research sample. Blood

(2010), while accepting that there is a correlation between increasing dependency and advanced old age, argues that it is not necessarily the case that those with significant support needs are in that age sector. In light of this, I have chosen to extend the lower age limit in order not to exclude those younger than in their eighties who may, chronologically speaking, not be typically classified as very dependent, but whose lived experience was nevertheless one of being considered to be old and incapable. And so, unless specifically defined otherwise, where I use the term elders, I use it to refer to those elders dependent on formal care support, rather than of a particular age within the old-age spectrum.

Empowerment: Where I refer to empowerment, I take on board the premise that power is not a 'zero-sum' concept (Tew 2002). That is, the argument that power relations cannot be well explained by the premise that there exists a finite amount of power which can be 'given' by those who possess it to those who do not (Morriss 2002). As Thompson (2007) comments: 'Power is not something that people can possess in a physical, tangible way. It is, to a large extent, about dispositions, institutionalised patterns, relationships and structures and, in order to appreciate the complexities of power and empowerment, we have to look at matters in a much more sophisticated way than just seeing empowerment as 'giving away power' (p. 60). I use the term 'empowerment' to reflect how raising the profile of reciprocity in the lives of dependent older people has the potential to raise the profile of older people's agency in general—that is, to suggest that their agency has been masked by ageist ideologies which work to normalise the assumption that they defer to others (Butler 1987; Bytheway and Johnson 1990; Bytheway 2011). I use the term, therefore, not to suggest that power be *given* to dependent older people but in the sense that they be helped to 'reclaim' their agency, in terms of both (a) their own biographical continuity as valued people and (b) with reference to the part they can play in changing power relations more broadly by challenging negative stereotypes through their competence to reciprocate.

Formal care: Where I refer to formal care, I use it to describe care support that is commodified (Ungerson 2000; Garey et al. 2002)—that is, it is support provided as a service by paid care workers, rather than that provided through kinship or friendship obligation or desire. I expand on this in Chap. 2 with reference to differing understandings of ageing and eldercare policy.

Gerontology: I use the term as shorthand for social and critical gerontology—social in the sense of an approach that looks beyond the individual ageing process as experienced in an atomistic way, and critical in the sense of incorporating a challenge to the disenfranchisement and marginalisation of older people (Phillipson and Walker 1987; Bernard and Scharf 2007; Townsend 2007). As Powell (2006) describes it: 'Critical gerontology is still concerned with structural inequalities, but is interested in moral concepts; it has a commitment not only to understanding marginality but also to challenging it' (p. 52).

Logocentricity: By adopting a phenomenological approach to this study I necessarily challenge logocentrism, in that I understand logocentrism to be a claim to knowing what is assumed to be *the* truth about any issue under scrutiny.

Logocentrism assumes the existence of authoritative and indisputable 'versions' of reality and so cannot account for differing perspectives, or the power to promote any one version as more 'true' than any other (Fox 1993).

Reciprocity: I use the term reciprocity to describe a general sense of 'usefulness'—be that to other individuals, communities or society in general—that is, a general intention to give as well as to receive in life, rather than an expectation of mutual exchange in equal measure in specific situations. While the term 'social engagement' may also describe the sense of usefulness to which I refer, I would suggest that the term 'reciprocity' incorporates a specific sense of 'giving back' that the former may not. Furthermore, it has implications for spiritual well-being which social engagement *per se* may not, in that it can be said to incorporate a sense of being valued for what one can give through that engagement. Social engagement *may* arise from a desire to 'give something back' but I would suggest that the two terms are not interchangeable, and I have chosen not to refer to the specific body of research on social engagement for that reason.

My focus on, and understanding of, reciprocity is influenced by the premise that quality of life is key to well-being (Walker 2005; Jordan 2007). In his model of 'the interpersonal economy' Jordan highlights the benefits of relationships which promote belonging, mutual interest and respect: 'People create *value* in their interactions. Positive interactions, which generate such emotions as closeness, respect, and belonging, are the main components in high levels of subjective well-being' (p. 12). Furthermore, he argues, they are integral to social well-being:

> Relationships with neighbours and other associates, schools and workmates and other fellow citizens are also significant for well-being. Interactions with others are the means by which people gain or lose in terms of the value which makes up well-being. Hence work which influences these relationships is significant for empowerment and inclusion of citizens, and for equality and justice in society (ibid).

It is these senses of *value* and *respect* which inform my understanding of reciprocity as an element of dignity (Nordenfelt 2003) and, as such, a concept that goes beyond mere engagement, because it is in the opportunity to reciprocate for help received that respect can be maintained, and self-esteem salvaged in the face of ageist assumptions about worthlessness. Giving is an important part of identity and well-being for everyone, but in situations of dependency as a result of care needs, there is a double effect: receiving care can be said to intensify the desire to give back (Lustbader 1991) while simultaneously denying the opportunity for reciprocity because of the common tendency, as I suggest later, for care providers to focus on providing care rather than promoting empowerment. It is (a) the low levels of awareness of the benefits of support based on mutual exchange (Bowers et al. 2011) and (b) the neglect of opportunities for dependent older people to feel valued and respected through being able to have, as Lustbader describes it, a 'counterbalance to their dependency' (p. 29), that underpins my concern that a lack of recognition of the significance of reciprocity in the care of older people may have negative implications for their spiritual well-being, as defined below.

Spirituality: Marx's work on dialectical materialism (Marx and Engels 1968) arose in response to Hegel's understanding of the dialectic of ideas (that is, the realm of mind or spirit) as the driving force for social change. As Tew (2002) explains:

> Where power is deployed in order to enforce aspects of the modernist status quo, its very application may be seen to generate the possibility of antagonism and resistance. The transformative potential of such antagonisms was first theorised by Marx in his reworking of Hegelian dialectic: instead of a focus on the generative potential of contradiction at the level of intellectual ideas and propositions, Marx relocated this discussion in terms of the contradictory interests of social classes at the level of material relations. (p. 182)

For Marx, class conflict was assumed to be the main force that drives social change, such that material, rather than mental or spiritual forces (Hegel used the German term, Geist, which can be translated as mind or spirit—Hegel 1977) were seen to underpin praxis. Some might therefore understand my focus on spirituality to be a backward step in the theorising of reciprocity—a return to pre-Marxist ideas and a rebuke of dialectical materialism. However, it is not my intention to challenge dialectical materialism but to suggest that a focus on spirituality can be complementary to it, as is evident in Sartre's later work on the integration of phenomenological concerns (emphasising meaning making) with Marx's focus on material conditions (Sartre 1963).

My understanding of spirituality is consistent with an emergent broad focus on spirituality as a search for meaning, and with a more particular focus on spiritual well-being through connectedness and social affirmation. This makes it a legitimate focus for social scientific, as well as theological, interest, as is evident from the growing body of literature which frames it as a social science concept (Canda and Furman 1999; Mowat 2004; Moss 2005; Thompson 2010; Coleman 2011). As Coleman (2011) comments:

> The challenges from existential questions at all crisis points throughout life find their culmination in old age. They are not of limited 'academic' interest. How older people actually find answers to these questions of meaning and purpose so as to sustain their daily lives are important topics for social research, and have huge implications for provision for those living in states of frailty and debility. (p.159)

The spiritual dimension of people's lives, and their search for meaning and purpose, is well expressed by Moss (2005) as follows:

> Spirituality, in other words, is a sort of 'shorthand' way of asking the fundamental questions about ourselves—what makes us 'tick'; what is important to us; what gives us a sense of meaning and purpose in our lives. In short, it asks of people what is their world-view. (p. 12)

And, building on the work of Canda and Furman (1999), he makes the point that the concept of a world-view necessarily incorporates connectedness with others. It is this sense of questioning how we fit into the world and who we think we are, would like to remain or would wish to become, that informs my use of the term spirituality in this book as something that incorporates, but is more than, mental and emotional well-being, and therefore requires a more sociological focus if it is to be well understood in old age. In terms of my own focus on reciprocity as an aspect of spirituality, the following personal reflection by the thanatologist, Jack Morgan (Morgan 2000, cited in Smith 2004) is particularly apt:

'For me, spirituality is a quest for meaning and an opportunity to make one's life important' (p. 50).

Strengths perspective: This refers to a helping approach not restricted to, but promoted in the social work field chiefly through the work of Saleeby (2008). It focuses not on pathology, but on working in partnership with service users to help them to identify and use their own strengths to resolve problems. As such, it can be seen to have the potential to challenge the pathologising of dependent older people as problems, though, as is discussed later, that potential may not be realised.

References

Bernard, M., & Scharf, T. (2007). *Critical perspectives on ageing societies* (eds.). Bristol: The Policy Press.
Blood, I. (2010). *Older people with high support needs: how can we empower them to enjoy a better life*. York: Joseph Rowntree Foundation.
Bond, J., & Cabrero, G. R. (2007). *Health and dependency in later life*, In: Bond et al.
Bond, J., Peace, S., Dittman-Kohli, F., & Westerhof, G. (2007). *Ageing in society: European perspectives on gerontology* (3rd ed.). London: Sage.
Bowers, H., et al. (2011). *Not a one-way street: research into older people's experiences of support based on mutuality and reciprocity*. York: Joseph Rowntree Foundation.
Butler, R. N. (1987). *'Ageism', in the encyclopedia of aging* 22–23. New York: Springer.
Byetheway, B. (2011). *Unmasking age: the significance of age for social research*. Bristol: The Policy Press.
Byetheway, B., & Johnson, J. (1990). 'On defining ageism'. *Critical Social Policy*, 10(29), 27–39.
Canda, E., & Furman, L. (1999). *Spiritual diversity in social work practice: the heart of helping*. New York: The Free Press.
Coleman, P. G. (2011). *'Ageing and the future of belief'*, In: Coleman.
Coleman, P. G. (2011). *Belief and ageing: spiritual pathways in later life*. (ed.). Bristol: The Policy Press.
Crossley, N. (2005). *Key concepts in critical theory*. London: Sage.
Fine, M., & Glendenning, C. (2005). 'Dependence, independence or interdependence?: Revisiting the concepts of "care" and "dependency"'. *Ageing and Society*, 25(4), 601–621.
Fox, N. J. (1993). *Postmodernism, sociology and health*. Buckingham: Open University Press.
Fromm, E. (1955). *The sane society*. New York: Rinehart.
Garey, A. I., Hansen, K. V., Hertz, R., & Macdonald, C. (2002). 'Care and kinship: An introduction'. *Journal of Family Issues*, 23, 703–715.
de Gregario, S. (1987). *Social gerontology: new directions*. (ed.), London: Croom Helm.
Grenier, A. (2007). 'Constructions of frailty in the English language, care practice and the lived experience'. *Ageing and Society*, 27(3), 425–445.
Gilleard, C., & Higgs, P. (2005). *Contexts of ageing: class, cohort and community*. Cambridge: Polity Press.
Haber, D. (2009). 'Gerontology: adding an empowerment paradigm'. *Journal of Applied Gerontology*, 28, 283–295.
Harrington Meyer, M. (2000). *Care work: gender, class and the welfare state*. London: Routledge.
Hill, R. D. (2005). *Positive aging: a guide for mental health professionals and consumers*. London: W. W. Norton.
Jewell, A. (2004). *Ageing, spirituality and well-being*. (ed.). London: Jessica Kingsley Publishers.

Jordan, B. (2007). *Social work and well-being*. Lyme Regis: Russell House Publishing.
Lustbader, W. (1991). *Counting on kindness: the dilemmas of dependency*. London: The Free Press.
Lymbery, M. (2010). 'A new vision for adult social care? Continuities and change in the care of older people'. *Critical Social Policy*, *30*(5), 5–26.
Marx, K., Engels, F. (1968). *Selected works*. London: Lawrence and Wishart.
Morriss, P. (2002). *Power: A Philosophical Analysis*. (2nd ed.). Manchester: Manchester University Press.
Moss, B. (2005). *Religion and Spirituality*. Lyme Regis: Russell House Publishing.
Mowat, H. (2004) '*Successful ageing and the spiritual journey*', In: Jewell.
Nordenfelt, L. (2003). 'Dignity and the care of the elderly'. *Medicine, Healthcare and Philosophy*, 6(2), 103–110.
Phillipson, C., Bernard, M., & Strang, P. (1986). *Dependency and Interdependency in Old Age: Theoretical Perspectives and Policy Alternatives*. London: Croom Helm.
Phillipson, C., & Walker, A. (1987). '*The case for a critical gerontology*', In: de Gregario.
Powell, J. L. (2006). *Social theory and ageing*. Oxford: Rowman and Littlefield.
Saleeby, D. (2008). *The strengths perspective in social work practice*. (5th ed.). Rugby: Pearson Education.
Sartre, J-P. (1963). *Search for a method*. New York: Vintage.
Smith, H. I. (2004). *Grievers ask: answers to questions about death and loss*. Minneapolis: Augsburg Fortress.
Sunderland, J. (2004). *Gendered discourses*. Basingstoke: Macmillan.
Tew, J. (2002). *Social theory, power and practice*. Basingstoke: Macmillan.
Thompson, N. (2010). *Theorizing social work practice*. Basingstoke: Palgrave Macmillan.
Thompson, S., & Thompson, N. (2008). *The critically reflective practitioner*. Basingstoke: Palgrave Macmillan.
Townsend, P. (2007). '*Using human rights to defeat ageism: Dealing with policy-induced structured dependency*', In: Bernard and Scharf.
Ungerson, C. (2000). '*Cash in care*', In: Harrington Meyer.
Walker, A. (2005). *Understanding quality of life in old age*. (ed.), Maidenhead: Open University Press.
Wenger, G. C. (1986). '*What do dependency measures measure?*' In: Phillipson et al.

Acknowledgments

First and foremost, I wish to acknowledge and thank all of the participants whose insights have given this study value. I thank them for giving their time and for sharing what was in their hearts, even when this was not always easy. I also want to acknowledge and thank everyone who has given me the intellectual, moral and technical support I have needed, and valued, along my research journey. In particular, I thank Neil Thompson for the support he willingly gives in ways too numerous to mention and Anna Thompson for many things but, in particular, her technical support and her patience. I thank Lora Deva Prasanna, for helping me to get the research project underway in Chennai, and for being so caring and supportive, especially given that she is always so busy herself. Many thanks also to Janice Stern for inviting me to publish, and to Janice, Kathryn Hiler and Christina Rodriguez Tuballes for their encouragement and support in getting the book to its final format.

Finally, I thank Louise Hardwick and Jason Powell for reading my work at a number of stages and for their insights that have helped to make the finished work as coherent and cogent as I hope you will agree that it is. In particular, I thank Jason for stretching my understanding of the social theory of ageing in general, and of how my own work fits in that context. Thank you, Jason, for believing in both the importance of this book, and my ability to write it.

Contents

Part I Foundations of Understanding

1 A Matter of Dignity 3
   Introduction 3
   Outline 4
   Background and Rationale 6
   Research Aims 8
   Research Questions 9
   Conclusion 12
   References 12

2 Setting the Context 15
   Introduction 15
   Developments in the Theorising of Ageing 15
      Logocentric Theories of Ageing 16
      The Hermeneutic Trend in Theories of Ageing 18
   Meaning in Context 21
   The Theoretical Orientation for this Study 23
   The Social Significance of Old Age 23
   Policy Relating to Eldercare in the UK and India 25
      Policy Informing Eldercare in the UK 26
      Policy Informing Eldercare in India 27
   The Phenomenological Lens 28
   Conclusion 30
   References 30

3 Reciprocity and Old Age 35
   Introduction and Background 35
   Social Space 37
      The Psychosocial Impact of Social Space 38
      Social Capital, Relationships and Social Space 40
      Well-Being and Social Space 42
   Social Time 44

	The Psychosocial Impact of Social Time....................	44
	Social Capital, Relationships and Social Time	46
	Well-Being and Social Time	48
Meaning Making at the Level of the Personal/Spiritual		50
	The Psychosocial Impact of Meaning Making at the Level of the Personal/Spiritual	50
	Social Capital, Relationships and Meaning Making at the Level of the Personal/Spiritual...........................	52
	Well-Being and Meaning Making at the Level of the Personal/Spiritual...	53
Meaning Making at the Level of Discourse and Institutionalised Patterns of Power		54
	The Psychosocial Impact of Meaning Making at the Level of Discourse and Institutionalised Patterns of Power	55
	Social Capital, Relationships and Meaning Making at the Level of Discourse and Institutionalised Patterns of Power	57
	Well-Being and Meaning Making at the Level of Discourse and Institutionalised Patterns of Power	58
Conclusion..		59
References		60

Part II Hearing Their Voices

4 Research Design and Methods 69
    Introduction... 69
    The Topic: Why Reciprocity? 70
    Research Aims.. 71
    Research Focus 72
    Research Design 73
    Data Collection 73
        Why Interviews? 73
        Why an Indian Dimension? 75
    Data Analysis... 76
    Reflections on Power.. 77
    Research Methods 79
        Preparation for fieldwork 79
        Sample 80
        Pilot Study 86
        Interviews... 87
        Transcription 90
        Analytical Tool... 91
    Methodological Limitations 92
    Conclusion... 93
    References 93

5 Findings 97
Introduction 97
The Study 98
The Participants 98
The Themes 101
The Participants' Perspectives 102
Social Space 103
    Place-in-the-World: Self-Perception as a Valued
    Member of Society 103
    Perception by Others (now) as a Valued Member of Society 105
    Perception of Opportunities to Reciprocate 108
    Expression of the Desire to Reciprocate 111
    Relationships Between Geographical and Social Space:
    Barriers and Enabling Spaces 113
    Understanding and Expectations of Care Relationship 114
Social Time 116
    Future Dimension/Personal Growth 116
    Self-perception as Changing in Terms of Being a Valued Person ... 119
    Expectations of Old Age in India and the UK:
    Continuity and Change 120
    Intergenerationality 121
    Developments in Communication Technology: Opportunity
    and Barriers 122
Meaning Making at the Level of the Personal/Spiritual 123
    Have I Got Anything to Give? 123
    A Sense of Being Valued Rather than Being a Burden 124
    Dependency as Existential Crisis: Am I the Same Person
    or Someone Different Now that I Need Help? 125
    Giving: Duty or Pleasure? 127
Meaning Making at the Level of Shared Meanings and Institutionalised
    Patterns of Power 128
    Health Discourse: Old Age as Illness 129
    Social Construction of Old Age 129
    Discourses Around Dependency: Care Model as Overprotective ... 130
    Welfare Policy and Practices 131
Conclusion 132
References 132

Part III The Implications

6 The Significance of the Findings for the Spiritual Well-Being of Older People Dependent on Formal Care 137
Introduction 137
Section One 138

Building and Maintaining Relationships	139
Are Opportunities to 'Give Back' or Remain 'Useful' When in Receipt of Care Recognised as Such and Facilitated?	143
The Recognition of Physical and Ideological Boundaries as Potential Barriers to Reciprocity	146
Section Two	150
Neglecting Aspiration	150
Personal Values: Continuity and Change	155
The Significance of Age Cohort	157
Section Three	159
Neglect of the Spiritual	159
Health Crises as Existential Crises	161
Reciprocity as Deference: A Case of Self-Betrayal?	163
A Shared World-View?	164
Section Four	165
Bio-Medical Discourses	166
Disability Discourses	169
Governmentality Discourse	170
Citizenship Discourses	171
Conclusion	172
References	173

7 The Significance of the Findings for the Theorising of Old Age 179

Introduction	179
Coherence	179
Meaningfulness	183
Silenced Voices? Reciprocity and the Co-construction of Knowledge	187
Reciprocity and Resilience: Challenging the Stereotype of Deficit	189
Conclusion	192
References	193

8 The Significance of the Findings for Eldercare Practice 197

Introduction	197
Reciprocity	198
Spirituality	202
Meaning Making	204
Dignity	206
Conclusion	207
References	209

9 Their Journeys and Mine 211

Introduction	211
The Significance of Social Space, Social Time and Meaning Making	212
A Sensitising Framework	213
Reflections on the Research Journey	214

Implications of the Findings for Future Research.................... 215
 Conclusion... 216
 References... 218

Appendix 1... 221

Appendix 2... 223

Appendix 3... 225

Index.. 227

Part I
Foundations of Understanding

Chapter 1
A Matter of Dignity

Introduction

I begin by highlighting the essence of the two-part thesis on which this book is based. The first premise is that there is the potential for older people who are significantly dependent on others for their care to experience that care as spiritually diminishing if the significance of reciprocity in their lives, and the fact that they may have aspirations to continue operating as valued citizens into their futures, are not recognised by those who facilitate or provide that care. The second is that highlighting the dialectical relationships that operate across the personal, cultural and structural aspects of social life has the potential to build on psychologically grounded literature relating to reciprocity in eldercare to take our understanding of the phenomenon forward in a more sociologically grounded direction.

In light of this, this book documents an exploration of the significance that older people dependent on formal care support attach to reciprocity in their lives, and whether they perceive its importance to be recognised, and such reciprocity to be facilitated, by those who support them.

I argue that, for the older people themselves, potential positive outcomes arising from reciprocity for well-being include:

- enhanced self-esteem and spiritual well-being through (a) positive affirmation as valued people (Jewell 2004a, b) and (b) a broader recognition of their positive role as agents in social change through their challenging of negative stereotypes that portray older people as a burden to society (Thornton 1995; Hatton-Yeo 2006, Grenier and Hanley 2007; Bowers et al. 2011); and
- the opportunity to experience and benefit from the connectedness which helps to facilitate the building up of social capital (McMunn et al. 2009; Windle et al. 2011)—that is, to have supportive networks on which they can draw in times of need, but within which they can also contribute to the well-being of others (Bowers et al. 2011).

However, they are not the only potential beneficiaries of the promoting of reciprocity in dependent old age, insofar as we as a society also stand to gain from how a challenge to deficit models in the theorising of dependent old age has the potential to highlight dependent old age as a time of strength as well as, in some ways, deficit. If the former, rather than the latter, were to better inform the theorising of dependent old age, then the potential would exist for those dependent on formal care to be recognised as a *resource*, as well as a *draw* on resources (Brindle 2011). In light of this, I suggest that attention needs to be given to the existence of barriers to the promoting of reciprocity in the lives of those dependent on formal care (Anderson and Dabelko-Schoeny 2010; Bowers et al. 2001, 2011), and to how these can be overcome in the interests of neither wasting the skills and knowledge of talented and competent people, nor compromising their spiritual well-being.

In what follows of this chapter I give a flavour of:

- what the study is about;
- why I have been moved to undertake it;
- why I feel a phenomenological lens to be enlightening; and
- why the study has an Indian dimension.

Outline

As I explain in more detail below, this is not a comparative study *per se*. Rather, drawing on the narratives of elders experiencing care support in Chennai, southern India, has enriched the data generated by the interviewing of UK-based elders by introducing more variety in terms of personal narratives about reciprocity, and the cultural and structural contexts in which they were experienced. Hearing the perspectives of elders living in the somewhat differing cultural and structural contexts of the UK and India has generated findings which support the premise that PCS analysis (P—personal, C—cultural, S—structural) offers a useful framework for furthering our understanding of reciprocity in the lives of older people dependent on formal care in a more sociologically grounded way than is currently visible in the research field.

PCS analysis is not a new concept. Indeed, it has been very influential in social work (and quite influential in probation work, nursing, youth and community work and counselling) primarily through the work of Thompson (2011a) for well over a decade. However, I believe that drawing on PCS analysis to highlight the existence of two sets of dialectical relationships (with shared understandings of old age and reciprocity as a mediating focus), is innovative in the study of reciprocity in eldercare—especially when enriched by an Indian perspective on expectations of elders, and the context of the formal support they are provided with.

In focusing on the operating of a 'double dialectic' (2011a), which I discuss in more detail in Chap. 2, I argue that researching the particular phenomenon of reciprocity in eldercare has better explanatory power if explained in the context of what Sibeon (1996) describes as 'sensitising frameworks':

> Substantive theories aim to provide us with new empirical information, whereas sensitizing theoretical frameworks are intended to furnish general orientations or perspectives; they are intended to equip us with ways of thinking about the world (p. 4).

In drawing on PCS analysis, then, it is my intention to construct a metanarrative that sensitises us to three sets of issues and the ways in which they interact to help explain the impeding or promoting of a particular phenomenon, but *not* a metanarrative in the sense of a grand or deterministic theory. In a later work, Sibeon (2004) clarifies the distinction as follows:

> Sometimes the expression *sensitizing theory* is used in place of the term metatheory. Sensitizing or (meta-) can and should inform the construction of substantive theories, but we have seen that the two types are distinct. In the social sciences substantive theories aim to generate new empirical information about the social world, whereas meta or sensitizing theories and concepts are concerned with general ontological and epistemological understandings; metatheories and meta-concepts are designed to equip us with a general sense of the kinds of things that exist in the social world, and the ways of thinking about the question of how we might 'know' that world.

He then goes on to say that:

> Metatheoretical generalizations of this kind [here he draws on Giddens's work on structuration theory as an example], however, are not the same as universal 'grand' generalizations associated with reductionist substantive theories such as Marxism, rational choice theory, and radical feminism… (pp. 13–14).

I therefore propose PCS analysis as:

(1) a useful sensitising framework for helping to explain how the meanings that dependent older people attach to the presence, or absence, of reciprocity in their lives are influenced both by the ways in which the societies they live in are structured with reference to age, and how this reinforces (and is sustained by) what old age is understood to mean in those contexts; and
(2) providing an element of coherence, in so far as it incorporates both macro and micro analyses and the interactions between them.

Post-structuralist and phenomenological thinking (as, indeed, existential thinking with which the latter is closely associated) have been influenced by Nietzsche's concept of 'perspectivism' (Heller 1988) by which he denies the existence of an objective 'reality' because it is not possible to see it without doing so from a particular perspective. As such, they have in common an emphasis on meaning making (for example, in the emphasis placed by writers such as Foucault (1972) on language and discourse, and the rejection of logocentrism—the assumption that meanings are fixed).While there are commonalities, there are also differences between post-structuralism and phenomenological/existentialist thinking, but it is beyond the scope of this study to examine them. And while I draw on Foucault's ideas in particular, it is only where they overlap with my phenomenological perspective.

Furthermore, I propose that the phenomenological focus, by both raising the profile of dependent older people as co-producers of knowledge, and contributing to an ongoing challenge to deficit models of ageing, addresses a gap in gerontological theorising. In Chap. 2, where I briefly chart the development of theories of ageing, I make the point that gerontological social work operates at the interface between theory and practice, and therefore has appeal because of:

(a) my self-definition as a 'passionate scholar', in light of my commitment to research that is both personally meaningful and has the potential to contribute to social justice (Holstein and Minkler 2007; Heinrich 2010); and

(b) my commitment to critically reflective, and therefore informed, practice (Thompson and Thompson 2008).

However, the literature within this strand appears to focus largely on policy, and less is known about how those in receipt of care feel about, and make sense of, how they are being supported—whether it makes them feel empowered or diminished (Grenier and Hanley 2007). I propose, then, that the phenomenologically grounded conceptual framework I use goes some way towards addressing that gap in the eldercare knowledge base.

Although this study is theoretically oriented, given my commitment to the integration of theory and practice (Thompson and Thompson 2008) I propose that my conceptual framework (visually represented in Fig. 1.1) provides an accessible tool for eldercare practitioners to better explore and understand the significance of both reciprocity in particular, and spirituality in general, in the lives of those elders who become dependent on them for formal care support.

I discuss the concepts more fully in the later chapters.

Background and Rationale

As I suggest in the later chapters, issues relating to demography (Wilson 2000; Timonen 2008; Kulshrestha 2009) and the inadequate funding of eldercare in both India and the UK (Chakraborti 2004; Dean 2009) have the potential to:

Fig. 1.1 Four-part phenomenological framework

(1) undermine the re-conceptualising of old age as a positive and affirming phase of life; and
(2) further polarise old age into a younger and more active '3rd age' and a '4th age' more typically characterised by, and conceptualised as, stagnation and decline (Powell 2006; Bond and Cabrero 2007; Gilleard and Higgs 2010).

An ongoing political agenda in the UK to address the neglect of dignity in eldercare (Social Care Advisory Group of the National End of Life Programme 2010; Care Quality Commission 2011) has the potential to challenge the depersonalising of older people through practices which neglect their humanity (in its sense as human-ness). So, too, has the promoting of dignity inherent in the Indian government's National Policy on Older Persons (Government of India 1999), as is discussed later. However, it is my concern that the significance of reciprocity as part of the dignity agenda may go unrecognised if it is not understood that having one's worth devalued or dismissed, through not having opportunities to do things with and for other people, has the potential to be spiritually diminishing and therefore an affront to dignity. Crossley (1996) makes this point in relation to intersubjectivity being at the heart of our understanding of the social world, in that people are constantly being influenced by their interactions with others in the construction of their own self-image and self-worth:

> self-consciousness is an intersubjective phenomenon … achievable only through mutual recognition between consciousnesses. Furthermore, our sense of self-esteem, pride and dignity is integral to this. These are feelings that we can only have relative to the other and they are thus bound to our relations and interactions with others (p. 17).

I came to this research primarily from a social work practitioner and educator perspective, though it builds also on an academic interest in the relationship between ageism and user involvement (Thompson 1997) and on loss in old age (Thompson 2002, 2007). From my observations as a nurse, social worker, practice teacher and educator I have noted a significant tendency on the part of assessors across the spectrum of the caring professions to neglect the spiritual dimension when assessing the needs of dependent older people, particularly in terms of the need to feel that they can still connect with the world and continue to give in some way as a counterpart to their being defined by others only, or primarily, with reference to decline and deficit (Palmore 2000; Tulle-Winton 2000; Gulette 2003; Powell 2006).

I have also noted that, despite:

(a) an increase in the involvement of older people in research activity (Peace 2002; Ray 2007);
(b) the efforts of the user involvement movement to highlight partnership working between providers and recipients of care (Kemshall and Littlechild 2000; Iliffe et al. 2010); and
(c) the existence of initiatives which request and respect the perspectives of older service users on the planning and delivery of support (Hayden and Boaz 2000),

the voice of older people with high support needs is still weak (Blood 2010) and older people's lived experience still tends to be considered epistemologically as inferior knowledge in relation to the perspectives of academics and eldercare professionals (Powell 2002; Preston-Shoot 2007). From these observations I have become particularly interested to:

- hear older people's stories in order to better understand whether they reflect my own interpretation that their perspective is undervalued; and
- analyse those stories to further my understanding of the extent to which shared perceptions of what it means to be old have an impact on expectations of reciprocity when older people's lives become significantly characterised by dependency.

While giving voice to the lived experience of elders receiving formal care in the UK alone would have been instructive in terms of knowledge development in this field, having the opportunity to hear about the desire and opportunity for reciprocity experienced by elders in Chennai, India, offered an even richer vein of experience to explore. I do not consider a fully comparative study between UK and Indian elders to have been feasible, given that I would not have been able to compare like with like in terms of eldercare provision. And so, while the context of changing demographics and social policy in India (Irudaya 2000) continues to provide fertile ground for research, it is the attempts by the older people themselves to make sense of their lives within those changing contexts that inspired me to draw them into the study in order to enrich the pool of data from which I might draw enlightening observations. As such, then, the Indian dimension is just that: an added dimension rather than the basis for direct comparison.

Research Aims

Building on some of the points made above, my research aims are fourfold:

1. To gain a service user perspective on whether older people dependent on formal care consider themselves to be capable of giving as well as receiving in life, and whether they perceive others as seeing them in that light. That is, to be privy to *their* meaning making, unmediated by the perceptions of others.
2. To analyse the narratives produced through interview with regard to similarities and differences of desire and opportunity, and how these relate to differences in the conceptualising of ageing.
3. To theorise the data by using a phenomenologically grounded conceptual framework which:
    (a) raises the profile of narrative perspectives;
    (b) highlights the significance of meaning making in people's lives and its relative neglect in studies of reciprocity and eldercare;
    (c) because it incorporates temporality, provides a platform from which to explore a future orientation in the lives of dependent older people, thereby highlighting the potential for a radical shift in how they are conceptualised; and
    (d) further develops a better sociological understanding of this phenomenon.

4. To propose the conceptual framework as an accessible theoretical tool for helping those involved in eldercare practice to:

    (a) appreciate the significance of reciprocity in the lives of the older people with whom they work;
    (b) understand its implications for their spiritual well-being; and
    (c) devise strategies for facilitating opportunities for the promoting or enhancing of reciprocity in eldercare.

Research Questions

The following questions have been designed to ground the research in the already identified phenomenologically significant concepts that relate to (a) making sense of social life, and (b) appreciating that meaning making operates at more than one level.

1. In what ways is the concept of social space significant in accounting for how reciprocity in the care of dependent older people in receipt of formal care is promoted or impeded?

This relates to the arenas in which relationships operate, and personal and social identities are constructed with reference to feedback acquired through connectedness with others (Biggs 1999; Crossley 1996). Barker (2008) comments that space is relationally defined, and draws on the work of Massey (1994) to suggest that social life is dynamic, in the sense that it is 'constituted by changing social relationships' (p. 376). It is not enough, however, to consider that dynamism in terms of social space alone. Dependent older people do not live their lives in a spatial vacuum, but nor do they live in a timeless one. In highlighting the significance of temporality for an understanding of the part reciprocity plays in life when people are dependent on others, I am influenced particularly by Heidegger and Giddens, both of whom refer to the interconnectedness of time and space. In his seminal work *Being and Time* Heidegger (1962) in studying the question of being, refers to human being and human beings as 'Dasein' (from the German verb 'to exist'—literally 'there-being') because it is as human beings that we have the capacity to ask questions about our existence.

Heidegger's thesis on being and time is complex and I do not intend to attempt to explore it in any depth here. Rather, I draw briefly on three of his core concepts which support my choosing of a phenomenological mode of enquiry, and the premise that both space and time have significance for how the older people in this study made sense of their experiences:

'*being-in-the world*'—this refers to his proposition that existence cannot be separated from our experience of it in everyday life through our interaction with objects and other people. From this premise it follows that there is no world without human beings (Cohn 2002). Sherratt (2006) makes a similar point:

> Heidegger goes as far as to say that we as self-interpreting beings are just what we make of ourselves in everyday life… not only are we able to understand—not only is understanding a crucial feature of our existence —but it *is* our existence (p. 80).

If we and the world are inseparable, and our experience of it is unique, it would seem that this highlights the relevance of meaning making for an understanding of social life.

'being-with others'—this relates to the understanding that all activity is connected to others, even if we are alone, because we are all involved in constructing the world in which we live. Inwood (1997) interprets Heidegger's concept as: 'Dasein alone is incomplete ... Dasein's world is essentially a public world, accessible to others as well as itself' (p. 40).

'thrownness'—this incorporates the understanding that our experience of being-in-the world has a temporal dimension to it. In describing existence as 'thrownness' Heidegger appears to be making the point that we have no choice about when and where we enter the world—we are 'thrown into' existence and have to accept that we will be subject to constraints over which we have no control. In acknowledgement of this, life is assumed to be a struggle to reach one's aspired-to future within those constraints. In recognising the interconnections between the future, present and past, Heidegger's concept of 'thrownness' chimes with those of 'intentionality' and the 'regressive-progressive method' in Sartre's work (Sartre 1963). With reference to intentionality, Thompson (1992) explains:

> This is a term which denotes that human actions are carried out with goals in mind. Actions are geared towards the future, towards achieving an aim or aims. Behaviourism, by contrast, stresses the influence of prior conditioning on behaviour. In other words, the past is seen as the most influential temporal dimension. In phenomenological terms, it is the future which is more important. Sartre maintains that consciousness is 'intentional'... it is this intentionality which makes life meaningful (p. 40).

In referring to the 'progressive-regressive method', Thompson (1992), drawing on the work of Sartre (1963), discusses the interplay between the regressive dimension (taking account of the past) and the progressive dimension (taking account of intention, ambition and so on). The latter constitutes the 'existential project' (Thompson, ibid).

This concept of 'project' in Heidegger's work is explained well by Frede (1993):

> Everything we are dealing with finds its meaning within this projection, and things have a meaning only insofar as they form part of it. Within this 'project' we make of ourselves, everything has its meaning and thereby its *being*. The design is, as the term suggests, directed into the future: we project ourselves into an anticipated future as the ultimate aim of our endeavours. But this is not the only temporal dimension that is at work in our projection, because our projection is not a free choice of the future. According to Heidegger, we cannot make any such projections without an existing understanding of the world and ourselves within it, an understanding determined by the past we have been and still are. Therefore, not only do we carry our past with us, as one carries weighty memories, but we always already understand ourselves and our projects in terms of the past and out of the past ... By temporality, he does not mean that we are, as are all other things, confined to time, nor that we have a sense of time, but rather that we exist as three temporal dimensions at once: it is being ahead of ourselves in the future, drawing on our past, while being concerned with the present that constitutes our being. (p. 64).

The significance of temporality in a sociological sense is echoed in Giddens's work, as evident in the following excerpt from Cassell (1993):

> Giddens' insight is that time and space are not just topics worthy of consideration by those interested in the social sciences. What he is arguing is much more radical: that excluding time and space from social analysis, or a priori one above the other, seriously distorts our understanding of the way social reality is constituted (p. 17).

The proposition that dependent older people make sense of their lives within a temporal context, as well as with regard to the spaces they live in, and the relationships they have with others, therefore informs the second of my research questions:

2. In what ways is the concept of social time significant in accounting for how reciprocity in the care of dependent older people in receipt of formal care is promoted or impeded?

This lays the foundation for, amongst other things, an exploration of such temporal issues as the recognition of existential projects to maintain a valued place in the world as one ages, and of changing attitudes towards, and social expectations, of old age.

3. In what ways is the concept of meaning making, as it relates to individuals and their spirituality, significant in accounting for how reciprocity in the care of dependent older people in receipt of formal care is promoted or impeded?

This third question guides the research to question what being able to maintain a valued place in the world—a sense of 'usefulness'—means to individuals, and whether this is recognised as significant by those providing care support when those people become defined by their dependency on others.

4. In what ways is the concept of meaning making, as it relates to discourses and institutionalised patterns of power, significant in accounting for how reciprocity in the care of dependent older people in receipt of formal care is promoted or impeded?

The final research question reflects that consideration needs to be paid to factors beyond individual concerns if we are to better understand the significance of reciprocity in eldercare. It serves to explicitly introduce the concept of power into the study, highlighting that individual aspiration can be facilitated or constrained by external factors relating to social structure (Giddens 2009; Stones 2005), and discourses that operate to normalise particular ways of living and working (Foucault 1977; Powell 2006).

These questions informed the construction of an interview schedule (Appendix 1) which, in turn, gave direction to my discussions with 14 elders, who are introduced in Chap. 4, where I discuss research methods. As the opportunity arose to interview individuals in Chennai before those I had identified in the UK, I was able to ensure that they met the same sample criteria, though I reiterate that I do not claim this as a comparative study.

I acknowledge that this study is broad in its scope but I would emphasise that, as an exploration of the factors that impede or promote reciprocity in eldercare, the purpose is partly to help map out the territory for future phenomenologically-based research in this field. I would argue, therefore, that its breadth is one of its strengths.

Conclusion

In this introductory chapter, I have accounted for how drawing on PCS analysis to explore the relationship between (a) how UK and India-based elders make sense of their experience of reciprocity while in receipt of eldercare (b) prevailing discourses about old age and dependency in both settings, and (c) the significance attached in each to age as a social division, transcends psychologically grounded studies to provide a more sociological dimension to studies of reciprocity in eldercare.

In addition to setting the scene for the innovation of applying PCS analysis to the study of reciprocity in the lives of older people dependent on formal care, I have incorporated the insights offered by it into a second analytical framework which I believe to be innovative in this context. I have done so in order to highlight that a sociologically and phenomenologically grounded understanding of reciprocity needs to be informed by:

(a) the significance of both spatial and temporal dimensions; and
(b) the understanding that meaning making operates not only at the level of personal spirituality, but also at the level of discourse, and in relation to the extent to which age is considered to be a valid basis for social division.

In the context of a strong focus (in the UK research context at least) on positive ageing, I have suggested that a focus on the meaning that reciprocity holds for dependent older people has the potential to challenge deficit models of ageing by highlighting that (a) dependency need not preclude reciprocity, and (b) dependent older people's meaning making has epistemological value.

Having provided an overview of the project, my rationale for both undertaking it at all, and for my choice of methodological approach in doing so, I now move in Chap. 2 to locate it in a broad range of extant literature relating to phenomenology, old age and eldercare policy.

References

Anderson, K. A., & Dabelko-Schoeny, H. J. (2010). Civic engagement for nursing home residents: a call for social work action. *Journal of Gerontological Social Work, 53*, 270–282.
Barker, C. (2008). *Cultural studies: Theory and practice* (3rd ed.). London: Sage.
Bernard, M., & Scharf, T. (Eds.), (2007). *Critical perspectives on ageing societies*. Bristol: The Policy Press.
Biggs, S. (1999). *The mature imagination: Dynamics of identity in midlife and beyond*. Buckingham: Open University Press.
Blood, I. (2010). *Older people with high support needs: How can we empower them to enjoy a better life?*. York: Joseph Rowntree Foundation.
Bond, J., & Cabrero, G. R. (2007) Health and dependency in later life, In J. Bond et al (Eds.), *Ageing and Society* (pp. 113–141). London: Sage Publications.
Bond, J., Peace, S., Dittman-Kohli, F., & Westerhof, G. (Eds.), (2007). *Ageing in society: European perspectives on gerontology* (3rd ed.). London: Sage.
Bowers, B.J., Fibich, B., & Jacobson, N. (2001). Care-as-service, Care-as-relating, care-as-comfort: Understanding nursing home residents definitions of quality. *The Gerontologist, 41*(4), 539–545.

References

Bowers, H., Mordey, M., Runnicles, D., Barker, S., Thomas, N., Wilkins, A., et al. (2011). *Not a one-way street: Research into older people's experiences of support based on mutuality and reciprocity*. York: Joseph Rowntree Foundation.

Brindle, D. (2011). Holding it together. *The Guardian*, 20th March, 2011.

Cann, P., & Dean, M. (Eds.), (2009). *Unequal ageing: The untold story of exclusion in old age*. Bristol: The Policy Press.

Care Quality Commission. (2011). *The state of healthcare and adult social care in England: An overview of key themes in Care 2009/10*. London: The Stationery Office.

Cassell, P. (Ed.), (1993). *The Giddens Reader*. Basingstoke: The Macmillan Press.

Chakraborti, R. D. (2004). *The greying of India: Population ageing in the context of Asia*. India: Sage.

Cohn, H. W. (2002). *Heidegger and the roots of existential therapy*. London: Continuum.

Crossley, N. (1996). *Intersubjectivity: The fabric of social becoming*. London: Sage.

Dean, M. (2009). How social age trumped social class. In Cann and Dean (Eds.), *Unequal Ageing: the untold story of exclusion in old age*. Bristol: The Policy Press.

Foucault, M. (1972). *The archaeology of knowledge*. New York: Pantheon.

Foucault, M. (1977). *Discipline and punish: The birth of the prison*. London: Allen Lane.

Frede, D. (1993). The question of being: Heidegger's project. In Guignon (Ed.), The Cambridge Companion to Heidegger. Cambridge University: Cambridge

Gilleard, C., & Higgs, P. (2010). Aging without agency: theorising the fourth age. *Aging and Mental Health, 14*(2), 121–128.

Government of India. (1999). *National policy on older persons*. New Delhi: India, Ministry of Social Justice and Empowerment.

Grenier, A., & Hanley, J. (2007). Older women and "frailty": Aged, gendered and embodied resistance. *Current Sociology, 55*, 211–228.

Guignon, C. (Ed.). (1993). *The Cambridge guide to Heidegger*. Cambridge: Cambridge University Press.

Gulette, M. M. (2003). From lifestory telling to age autobiography. *Journal of Aging Studies, 17*, 101–111.

Hancock, P., Hughes, B., Jagger, K., Paterson, R., Russell, E., Tulle-Winton, E., et al. (Eds.), (2000). *The body, culture and society: An introduction*. Buckingham: Open University Press.

Hatton-Yeo, A. (Ed.), (2006). *Intergenerational programmes: An introduction and examples of practice*. Stoke-on-Trent: Beth Johnson Foundation.

Hayden, C., & Boaz, A. (2000). *Making a difference, better government for older people evaluation report, coventry*. UK: Local Government Centre, University of Warwick.

Heidegger, M. (1962 trans.). *Being and time: A translation of Sein und Zeit*. New York: University of New York Press.

Heinrich, K. T. (2010). Passionate scholarship, 2001–2010: A vision for making academe safer for joyous risk-takers. *Advances in nursing science, 3*(1), E50–E64.

Heller, E. (1988). *The importance of Nietzsche*. London: The University of Michigan.

Holstein, M. B., & Minkler, M. (2007). Critical gerontology: reflections for the 21th Century. In Bernard and Scharf (Eds.), *Critical perspectives on ageing societies* (pp. 13–26). Bristol: The Policy Press.

Iliffe, S., Kharicha, K., Kharicha, D., Swift, C., Goodman, C., & Manthorpe, J. (2010). User involvement in the development of a health promotion technology for older people: Findings from the Swish project. *Social Care in the Community, 18*(2), 147–159.

Inwood, M. (1997). *Heidegger: A very short introduction*. Oxford: Oxford University Press.

Irudaya, R. S. (2000). Ageing in India: Retrospect and prospect. *Indian Social Science Review, 2*(1), 3–47.

Jamieson, A., & Victor, C. R. (Eds.). (2002). *Researching ageing and later life*. Buckingham: Open University Press.

Jewell, A. (2004a). Nourishing the inner being: A spiritual model. In A. Jewell (Ed.), *Ageing, Spirituality and Well-being* (pp. 11–26). London: Jessica Kingsley.

Jewell, A. (Ed.), (2004b). *Ageing, Spirituality and Well-being*. London: Jessica Kingsley Publishers.

Kemshall, H., & Littlechild, R. (Eds.), (2000). *User involvement and participation in social care: Research informing practice.* London: Jessica Kingsley.
Kulshrestha, S. (2009). Social security for elderly in India. www.globalaging.org/ruralageing/world/htm.
Massey, D. (1994). *Space, place and gender.* Cambridge: Polity Press.
McMunn, A., Nazroo, J., Wahrendorf, M., Breeze, E., & Zaninotto, P. (2009). Participation in socially-productive activities, reciprocity and well-being in later life: Baseline results in England. *Ageing and society, 29*, 765–782.
Palmore, E. (2000). Ageism in gerontological language. *Gerontologist, 40*(6), 645.
Peace, S. (2002). The role of older people in social research. In A. Jamieson & C. R. Victor (Eds.), *Researching Ageing and Later Life.* Buckingham: Open University Press.
Powell, J. (2002). The changing conditions of social work research. *British Journal of social work, 32*, 17–33.
Powell, J. L. (2006). *Social theory and ageing.* Oxford: Rowman and Littlefield.
Preston-Shoot, M. (2007). Whose lives and whose learning? *Evidence and Policy, 3*, 343–360.
Ray, M. (2007). Redressing the Balance? The participation of older people in research. In M. Bernard & T. Scharf (Eds.), *Critical Perspectives on Ageing Societies* (pp. 73–87). Bristol: The Policy Press.
Sartre, J.-P. (1963). *Search for a method.* New York: Vintage.
Sherratt, Y. (2006). *Continental philosophy of social science: Hermeneutics, genealogy, and critical theory from Greece to the twenty-first century.* Cambridge: Cambridge University Press.
Sibeon, R. (1996). *Contemporary sociology and policy analysis: The new sociology of public policy.* London: Kogan Page/Tudor.
Sibeon, R. (2004). *Rethinking social theory.* London: Sage.
Social care advisory group of the national end of life programme. (2010). *Supporting people to live and die well: A framework for social care at the end of life.* London: National Health Service.
Stones, R. (2005). *Structuration theory.* Basingstoke: Palgrave Macmillan.
Thompson, N. (1992). *Existentialism and social work.* Avebury: Aldershot.
Thompson, N. (Ed.), (2002a). *Loss and grief.* Basingstoke: Palgrave Macmillan.
Thompson, N. (2011). *Promoting equality: Working with diversity and difference* (3rd ed.). Basingstoke: Palgrave Macmillan.
Thompson, S. (1997). *User involvement: Giving older people a voice.* Wrexham: Prospects Publications.
Thompson, S. (2002). Older people. In N. Thompson (Ed.), *Loss and Grief: A guide for human services practitioners* (pp. 162–173). London: Palgrave Macmillan.
Thompson, S. (2007). Spirituality and old age. *Illness, crisis and loss, 15*(2), 169–181.
Thompson, S., & Thompson, N. (2008). *The critically reflective practitioner.* Basingstoke: Palgrave Macmillan.
Thornton, P. (1995). *Having a say in change.* York: Joseph Rowntree Foundation.
Timonen, V. (2008). *Ageing societies: a comparative introduction.* Maidenhead: Open University Press.
Tulle-Winton, E. (2000). Old bodies. In P. Hancock et al (Eds.), *The. body, culture and society* (pp. 64–83). Buckingham: Open University Press.
Wilson, G. (2000). *Understanding old age: Critical and global perspectives.* London: Sage.
Windle, K., Francis, J., & Coomber, C. (2011). *Preventing loneliness and social isolation: Interventions and outcomes, SCIE Research Briefing 39.* London: SCIE.

Chapter 2
Setting the Context

Introduction

As the lived experience of reciprocity in elder care would seem to be under-represented in the literature relating to the care of older people significantly dependent on others, I draw on a fairly wide range of literature to situate, and set the scene for, the study that follows.

In Chap. 3 I focus particularly on the literature relating to the concept of reciprocity itself. The strands of literature I explore there reflect four themes which have phenomenological significance in the lives of dependent older people because they relate to their validation and affirmation as valued citizens. But, while the research itself focuses on individual experience and narrative, I consider that to analyse it without exploring the broader contexts in which individuals make sense of their experience would run counter to my intention to add a more sociological dimension to the predominantly psychological focus that currently characterises studies in this field. Furthermore, to ignore epistemological questions about the value accorded to differing perspectives within the research dynamic, and the significance of the dynamic across personal, cultural and structural factors for a holistic understanding of ageing itself, would also reinforce the atomistic focus I seek to avoid. In this chapter, therefore, I briefly review the literature relating to:

- developments in the theorising of ageing;
- the social significance of age;
- policy informing eldercare in the UK and India; and
- the phenomenological lens.

Developments in the Theorising of Ageing

As discussed in the introductory chapter I seek not only to develop our understanding of reciprocity in old age, but also to contribute to developments in how ageing itself is theorised. This is not just for its own sake, but as part of the rationale for the study—that is, to provide a challenge to theories where the conceptual framework is based on deficit and decline.

In light of the vast amount of literature on which I could draw, and the restrictions I have in terms of space, I present just a broad overview of the literature relating to the theorising of ageing, in order to set my own theoretical orientation within it. While I recognise that these are very broad categories that do not address specific debates within and between them, for clarity of presentation for the purposes of this study, I distinguish between broadly logocentric and hermeneutic theories.

Logocentric Theories of Ageing

By logocentric theories of ageing, I mean those which are associated with modernism (Fox 1993), in the sense of being underpinned by an assumption that there exists one Truth, which can become evident if only that 'Truth' can be discovered. As Powell (2006) suggests, modernist theorists tend to theorise ageing from macro perspectives and in relation to vested interest. We can see how these are reflected in the following paradigms:

Functionalism

From this perspective, old age is seen as a problem for the smooth functioning of society. For example, Cumming and Henry (1961) explained the marginalised position of older people in terms of the expectation that they should 'disengage' from their social roles in anticipation of their eventual deaths, thereby lessening the disruptive effect on the smooth running of society and the implications for their own well-being. It is interesting to note that early Hindu philosophy describes a similar phase of disengagement (*Vanaprasthasrama*—the third stage of life) as a normative process within a cycle of birth, life, death and rebirth (Jacobs 2010) although its purpose seems to be as much to do with individual spirituality and fulfilment as it is to the smooth running of society, as is evident in the following comment from Chatterjee et al. (2008):

> [it] coincides with the stage of life when biological decay starts rapidly. Here the householder … transfers his [sic] responsibility to the younger generations and engages in the pursuit of higher knowledge and spiritual advancement. Needless to say, this stage of life is conceived with both individual and society in mind. It helps younger persons to shoulder responsibilities, thereby ruling out inter-generational conflict. At the individual level, it encourages disengagement from material pursuits and inclines the person towards self-realization. (p. 20)

While the stage described above arises from the ancient Vedic tradition, which is not necessarily taken on board in its totality in the more secularised regions and communities of modern India, the guiding principles of the Hindu faith nevertheless appear to continue to strongly influence the daily lives and aspirations of those who practise it—estimated at four-fifths of the total Indian population (Farndon 2007). Pragmatism, as is the case with many religions, co-exists with, and is said to dilute, the moral imperatives attached to Hinduism (Varma 2005). However, strong elements of adherence to what is accepted as one's duty in life (dharma), especially in terms of what is expected of people in or approaching old age, were reflected in the narratives of participants in the Indian sample of my study, as I discuss in later chapters.

This adoption of new roles, or a new focus in life, has parallels with *activity theory,* (Havighurst and Albrecht 1953). From this perspective social participation (albeit in different roles), rather than social disengagement, is seen as the key to personal and social well-being. While, on the face of it, it appears to reframe old age as a time of growth and opportunity, the diversity of lived experience is lost within its 'one-size-fits-all' approach and its focus on 'the interests of society'. What activity theory seems to have in common with disengagement theory is the assumption that old age is necessarily a time of deficit—that is, we have to recognise incapacity and limitations and accommodate them through lifestyle changes.

Political Economy

While functionalist theories of ageing analyse it in relation to the vested interest of society as a whole, proponents of the political economy approach focus on the role of political and economic systems in shaping how old age is experienced (Walker 1981; Phillipson 1982; Minkler and Estes 1998). For example, policies such as enforced retirement, which marginalise and disenfranchise older people by removing them from the sphere of active and productive employment, are said to work to the advantage of younger people who then have access to the best opportunities in the workplace. The relationship between the ageing worker and the economy is a complex and changing one, as we can see from current debates in the UK about retirement age and pensions policy (Jackson 2009). Indeed the argument that both young and old people necessarily constitute 'dependent' groups has been challenged (Arber and Ginn 1995). However, it is that very fluidity that those conceptualising it from a political economy perspective would cite as evidence that dependency in old age is socially constructed in line with the interests of the economy and the state (Townsend 2007). Again we see old age being conceptualised as a period of life characterised by deficit, decline and decreasing 'usefulness'.

Early Feminism

As a reaction to theorising that failed to take account of the specific experiences of women, the feminist movement looked to explaining their perceived inferiority and to challenge women's absent or marginalised position in the public sphere (de Beauvoir 1972; Rowbotham 1973). Within these discourses, the particular situation of older women was minimally addressed and often theorised in relation to the vested interests of patriarchy (de Beauvouir 1977). As May and Powell (2008) suggest, even where sex differences began to be theorised in terms of the social construction of difference, women still tended to be conceptualised as a homogeneous group and their experiences analysed in relation to 'foundations set by men' (p. 175).

Later developments within feminism have begun to recognise the uniqueness of both women and men as individuals within those categories, and that their life experiences as men and women are also about their experience of *ageing* as men and women (Fairhurst 2003; Moss and Moss 2007; Park et al. 2009; Stephens et al. 2011). As I discuss below, there are social expectations about age-appropriate roles and obligations, and these are gendered expectations. It is clear from demographic data that, in both India and the UK, there is a gender imbalance in the aged population, with women predominating in the age group featured in this research project (Wilson 2000; Timonen 2008). Whether this has any implication in relation to expectations of, or opportunities for, reciprocity in old age will be a feature of the data analysis in Part Three.

While these are very brief summaries which necessarily overlook many of the complexities contained in the arguments raised by commentators writing within those paradigms, my aim has been merely to make the point that the common ground for those writing from a broadly logocentric perspective on ageing is their nature as 'grand' theories. Grand theories have been critiqued by Sibeon (2004) as reductionist, in the sense that attempts are made to explain complex situations in ways that do not recognise that complexity. That is, they tend to focus on universality and universal themes and to conceptualise old age as an all-encompassing, undifferentiated concept, rather than as a dynamic stage of life (Blood 2010) stratified in a myriad of ways, including class, gender, ethnicity, sexuality and differential ability (Boneham 2002; Arber et al. 2003; Boerner and Rheinhardt 2003; Bagga 2008; Fredriksen-Goldsen and Muraco 2010). And, because logocentric theories tend to be informed by prevailing scientific and bio-medical discourses that associate ageing with physical decay, the conceptualisation of old age, or more particularly, advanced old age, as a period of decline is a dominant and enduring feature in this strand of literature (Powell 2006).

Furthermore, because of their structural focus, there is a tendency to neglect personal and cultural/discursive considerations. As I discuss later with reference to the dialectic between subjectivity and objectivity, this is problematic for an understanding of the experience of old age as unique, multifaceted and life affirming.

The Hermeneutic Trend in Theories of Ageing

Theories arising from within this paradigm have their foundation in the understanding that multiple 'truths' exist because accounts of social phenomena are shaped by differences in perception, and by the power relations that inform that meaning making. One of the features of postmodernist thought (influenced as it is by post-structuralism) is the rejection of 'logocentrism', the assumption that an unmediated knowledge of the world is possible, that we can have fully objective knowledge of the world and specific aspects of it. Positivism, with its emphasis on scientific rationality, is an example of logocentric thinking. The postmodernist view can be traced back to Nietzsche's 'perspectivism' (Heller 1988)—his rejection of the idea that there can be absolute truth, separate from subjective perception. This fits with Rorty's (1998) criticism of approaches which assume that there can be a 'view from nowhere'.

The rejection of logocentrism presents opportunities for criticisms of 'relativism'—the ideas that, if there is no fundamental truth, then any one perspective is as valid as any other. Rorty (1989) challenges this conceptualisation of truth, arguing that it needs to be understood in practical, situated terms and not in absolute terms. He therefore sees claims of relativism as oversimplifications of a very complex field of inquiry. Foucault (1972) offers a more complex understanding of such matters, arguing that what counts as 'true' depends on the discourse(s) in which such inquiry is being made. Discourses, for Foucault, are 'regimes of truth' that produce 'truth effects' (that is, they shape what is regarded as true or valid). So, from a Foucauldian view, claims of relativism are, as Rorty also argued, oversimplifications, because they neglect the power of discourses to make some perspectives appear more valid than others.

A discourse is, literally, a conversation, but in the work of Foucault it is extended to refer to forms of language and associated assumptions and behaviours which contribute to a particular world-view. In this respect, discourses are important aspects of meaning making and therefore have connections with phenomenology, insofar as both have their roots in a hermeneutical approach—that is, one which emphasises the importance of meaning. Because of this link, the concept of discourse is one that can be usefully employed as part of my theorising of ageing.

Characteristically, the literature in this paradigm focuses on local narratives and discourses, not necessarily to the exclusion of, but generally according less importance to, structural concerns. They reflect a recognition of diversity within older populations and their need to find their own way in the world rather than being able to rely on preconceived, fixed and normative rules. The tendency, then, is to focus on aspects of ageing, rather than on ageing as an overall concept. Research studies such as those focusing on differing perceptions towards older men and women as spousal carers (Rose and Bruce 1995); parent-daughter care relationships (Datta et al. 2003); and social engagement in relocated older adults (Dupuis-Blanchard et al. 2009) are typical of the trend toward micro-analysis. So too is a focus on the ageing body, as apparent in the work of Gilleard and Higgs (2005) who discuss the potential for new technologies to enhance quality of life in old age. The focus is not always on improved health or positive body image, however.

For example, the relationship between negative body image and ageing has attracted attention (Featherstone and Wernick 1995) and the conceptualising of care work as 'bodywork' (Twigg 2004) could be said to help perpetuate negative images of old age as being linked with physical decline. Nevertheless, a focus on old age as a positive time of life remains typical of recent and current theorising in the field of gerontology (Rowe and Kahn 1999; Hill 2005; Haber 2009). And while they would no doubt agree with the argument that older populations are still stratified by class, so that not all older people can participate equally (Cann and Dean 2009). Gilleard and Higgs (ibid) challenge the association between old age and decline by suggesting that, for the current cohort of older people (in western societies at least) old age can be experienced positively as a time characterised by increased opportunities for leisure and consumerism:

> We have shown that many of the issues surrounding ageing societies are ones of modernity's making: namely the rise in life expectancy, the fiscal crisis of pensions funding, the problems of urbanization and urban communities, and the changing nature of household and family. But they are also issues related to the agency of post-war cohorts. The transformation of later life from residual category to leisure culture emerged out of the rising affluence of those who rejected the view that saw utility only in labour. (p. 150)

Indeed, as we see below, theorising around affluence and opportunity can be seen to inform several policy strands which run through current policy in the UK.

There is, therefore, evidence in the gerontological literature of a change in the conceptualisation of old age from one associated with decline to one associated with opportunity and empowerment. For example, quality of life in old age is well researched (McKee et al. 2005; Walker and Hennessy 2004; Windle et al. 2011) as too are 'successful' or 'healthy' ageing (Franklin and Tate 2009; Hung et al. 2010) and competence in old age (Shachter-Shalomi and Miller 1997; Cohen 2005; Lamb et al. 2009). Furthermore, the existence of literature and initiatives that promote opportunities for learning (Hafford-Letchfield, 2010; University of the Third Age: Age and Opportunity); employment (Kulkarni 2009; O'Hara 2010); and sport (Tokarski 2004) also support the premise that old age is increasingly being viewed more as an extended holiday or time for investing in oneself, and less as the commonly used term, 'God's waiting room'.

I would agree that a focus on positive and active ageing has the potential to reframe old age as a time of empowerment, opportunity and growth, although not for all to the same degree. Furthermore, I would suggest that literature which focuses on the deconstruction of ageing, and the meanings attached to it, has succeeded in;

- raising the profile of positive ageing;
- addressing diversity within ageing populations; and
- highlighting the neglect of micro-analysis

all of which are to be welcomed. However, where recent literature has been successful in moving attention away from old age as a time of decay and decline, for the most part the focus has been on those classified as 'old' on chronological grounds but from the younger and more active end of the age spectrum—commonly described as the 'third age' (Laslett 1996). Literature which reflects a positive reframing of life in the 'fourth age'—described by Marcoen et al. (2007) as: 'A time of increasing disability, frailty and constriction of opportunity' (p. 60) and by Gilleard and Higgs (2010) as: 'a kind of terminal destination—a location stripped of the social and structural capital that is most valued and which allows for the articulation of choice, autonomy, self-expression, and pleasure in later life' (p. 123) is, as I suggest later, much thinner on the ground.

Strong arguments have already been made for race, gender and disability to be understood as social constructions—the products of social and cultural practices—rather than part of a 'natural order' (Butler 1990; Pilkington 2003; Barnes and Mercer 2010). Such arguments have been influential in their respective areas of theory development, as too (though perhaps to a lesser extent) has the argument that old age be considered a social construction (Townsend 2007; Bytheway 2011). However, with specific regard to the 'fourth age' as a dimension of old age, the biomedically influenced 'decline as natural' premise is proving harder to displace. By demonstrating that dependency need not preclude 'usefulness' and personal growth, I suggest that we should recognise the 'fourth age' to be a social construction too.

Meaning in Context

Underpinning my decision to draw on PCS analysis to inform this study is the recognition that it is grounded in meaning making and therefore consistent with an exploration of spirituality—meaning making related to a sense of purpose—that is experienced at a personal level but not atomistically (Moss and Thompson 2007; Holloway and Moss 2010). With reference to the links between spirituality and equality agendas Moss and Thompson (2007) comment as follows:

> Far from being a marginalised or marginalizing concept, spirituality is taking us into the very heart of the contemporary debate about equality and diversity, with its ways in which the celebration of individual uniqueness and diversity can enhance a community's richness… the importance of equality issues for spirituality is evidenced by the fact that inequality and discrimination can clearly stand in the way of spiritual fulfillment. (p. 6)

Given that I seek to challenge atomistic explanations of reciprocity, and to highlight the power of ageism to downplay its significance for spiritual well-being in old age, PCS analysis provides a suitably sophisticated framework for setting individual meaning making in context, and I therefore now turn to explaining it more fully.

Psychological theories in general have attracted criticism from a sociological point of view for their tendency to focus on the individual and to neglect wider social concerns (Rubenstein 2001; Stones 2005; Thompson 2010). By contrast, structurally based sociological theories (some forms of Marxism, for example) have been criticised for focusing on wider structures without adequately theorising human agency. Giddens (2009) went some way towards avoiding both of these flaws by developing structuration theory in which he emphasises the importance of the dialectical interaction between structure and agency.

Other theorists have gone in a different direction. Post-structuralist thinkers, largely but not exclusively influenced by Foucault (Rouse 2003), have emphasised the significance of discourses as frameworks of language, meaning and power and have paid relatively little attention to structure or agency. In some ways, then, post-structuralism addresses some of the important issues not covered by structuration theory, but lacks the strength of the latter in terms of being able to explain the significance of the interaction of personal agency and wider social structures.

Thompson (2010) draws on existentialist thought to propose an analytical framework that encompasses the social field more holistically, drawing on the strengths of post-structuralism and structuration theory without their respective flaws. PCS analysis (introduced in Thompson 1993, but discussed in more detail and further developed in Thompson, 2010, 2011a) attempts to capture the complexities involved by illustrating the interaction of three levels: personal, cultural and structural. The personal and structural levels are similar to agency and structure in structuration theory, with the cultural level (the level of shared meanings, taken-for-granted assumptions and 'unwritten rules') having much in common with the idea of discourses as frameworks of meaning and thus power. In Thompson (2011a) PCS analysis is presented in terms of a double dialectic (see Fig. 2.1).

The first dialectic is that between the personal level and the cultural level: the individual is influenced by discourses/frameworks of meaning but is not simply determined by them; he or she also contributes to the cultural level by, for the most part, reinforcing it and perhaps occasionally challenging it. The cultural and the structural levels then form the second dialectic. Cultural assumptions (about age-based norms, for example) both reflect and reinforce the legitimacy of structuring society using age as a basis for social division—the cultural and structural levels influence and reinforce each other. This conceptualisation goes beyond structuration theory by showing that agency and structure are mediated by culture, that is, by institutionalised meanings. It also goes beyond post-structuralism by bypassing criticisms that agency and structure are neglected.

This more holistic approach, incorporating agency, culture/discourse and structure, provides a framework for analysing the factors that impede or promote reciprocity. What is particularly helpful is the central element of culture which is premised on meanings. Building on Thompson's original figure, this can be represented diagrammatically as follows (see Fig. 2.2):

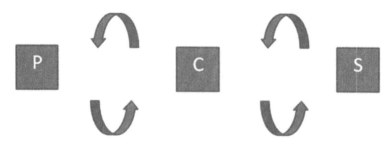

Fig. 2.1 The double dialectic

The Theoretical Orientation for this Study

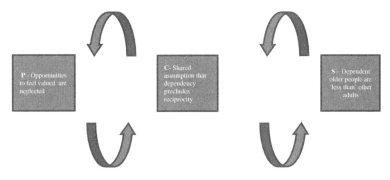

Fig. 2.2 The impeding of reciprocity

The Theoretical Orientation for this Study

In theorising reciprocity in the care of dependent older people I am therefore proposing an approach which;

- as a social constructionist perspective fits with Sibeon's notion of a 'sensitizing' theory as discussed above—that is, one which explores micro-processes without losing sight of the contexts in which they are played out;
- draws on the phenomenological paradigm because it has meaning making at its core. As such, it has the potential to account for differing perspectives on 'reality' without losing sight of structural concerns which:
    a) inform our understanding of the lived experience of reciprocity; and
    b) address the tendency in research on care relationships for the professional voice, or 'take' on reality to be prioritised over service user perspectives; and
- acknowledges that, as rooted in existentialist theorising (Thompson 2010), people change because of the constant interplay in their lives between subjectivity and objectivity. In highlighting the potential this offers for a greater recognition of a future orientation in the lives of older people, this chimes well with the potential for the phenomenologically grounded model to address a contested, but still enduring emphasis in gerontological literature (as it relates to dependent older people) on incapacity and decline (Palmore 2000; Powell 2006; Bond and Cabrero 2007), by focusing on the potential for continuing growth and personal fulfilment, and their contribution to well-being.

The Social Significance of Old Age

Informed as it is by meaning making at the level of discourse and institutionalised patterns of power, this research cannot be presented without locating it in the context of literature in which it is argued that chronological age is imbued with social significance. We can see that it is socially constructed because old age is neither universally defined nor constant across time—older people are venerated in some societies and the subject of negative stereotyping in others (Butler 1987; Blaikie and Hepworth 1997; Whitbourne and Sneed 2002). As I have commented in a previous work:

> Age provides a basis for social stratification or hierarchy, in a similar way to class, ethnicity, gender and sexual orientation and, as social significance and esteem are attached to particular life stages, it is clearly a sociological phenomenon as well as a psychological one. (Thompson, S. 2007, p. 170)

In a similar vein, Timonen (2008) refers to age as having a sociocultural dimension, which adds value judgements to the analysis:

> This refers to society's expectations of older people. Regardless of how older people view their bodies, minds and capabilities, society can place extensive limits on their ability to act in accordance with those views. All societies tend to assign older people certain 'age appropriate' roles, such as retirement and grandparenthood, and suitable patterns of activity or inactivity… (p. 8)

Situating a study of reciprocity in old age as experienced within two different cultural contexts (that is, the UK and India) therefore requires attention to the manner in which old age itself is constructed and discussed within those contexts. From a UK perspective the lives of older people in their 'fourth' age (Phillipson 1998; Gilleard and Higgs 2010) continue, in many respects, to be adversely affected by discriminatory attitudes and practices (Nelson 2005; Memmers 2004; Westerhoff and Tulle 2007), which reinforce the image of old people as 'past it' and strengthen the association between old age and decline (Palmore 2000; Tulle-Winton 2000).

In contradiction to this negative view of ageing, there is a common-sense view that older people are venerated in traditional societies because of their accumulated expertise and experience. This view is upheld by both Chakraborti (2004) and Chatterjee et al. (2008) as being true of Indian culture in general, although both agree that this veneration is being challenged in recent times because of the intergenerational conflict that is arising in the wake of rising numbers of older people needing support, and the falling number of younger relatives available and willing to care for them as part of extended families (Bagga 2008). It is clear from recent legislative change that the Indian national government is keen to promote family responsibility for the care of older people, in that The Senior Citizen Act 2007 allows for sanctions to be taken against those families who refuse to look after dependent older relatives. However, it may be that, as has reportedly happened with The National Policy on Older Persons, 1999, it will not be implemented or policed effectively at the level of individual states (Datta 2008).

Furthermore, the rise in the number of care homes being established in response to demand from older people seeking, or being forced into, institutionalised care (Liebig 2003; Srikrishna 2006) and in instances of abuse (Srinivas and Vijayalakshmi 2001) suggests that supporters of family based care cannot count on the moral imperatives that once strongly characterised Indian society.

It was interesting to note from discussions with Indian social work colleagues prior to the setting up of this research, that old age itself is not generally seen as an area of concern to be addressed by them, That is, older people only attract the attention of welfare agencies if their lives become problematic, such as in cases of abuse or abandonment, which are often conceptualised as problems of family dynamics rather than of old age. In what appears to be an otherwise comprehensive volume on social work and social welfare in India, Nagendra (2005) does not refer specifically to older people as a client group or social work speciality, mentioning them only briefly alongside references to 'the destitute and handicapped'—perhaps supporting the premise that old age is only seen as a problem when associated with a problem?

One further point of potential significance in terms of how old age is constructed in the two cultures under investigation is the relationship between mind and body. In Western cultures, bio-medical discourses are largely successful in constructing and reinforcing the message that old age is an illness or form of disability, rather than a stage of life that may be characterised by illness or disability but is not defined by it (Bond and Corner 2004). However, this association may be given less credence in India where a more holistic and cosmic interpretation of the life span seems to be prevalent (Jacobs 2010). In the introduction to Chatterjee et al. (2008), a collection of papers on palliative care, the editors make the following points:

> The prioritization of the body is almost absent in Indian philosophy. Each individual soul is immortal by nature and continues to exist even after the death of the body. Conceptually, a distinction is also made between gross and subtle body. The soul, having enjoyed certain pleasures and having fulfilled certain desires, leaves the gross body for another. The subtle body, on the other hand, accompanies the transmigrating spirit and is the essential link in the continuity of life because it is not destroyed by death. It is in this sense that death does not mean destruction or even reduction of anything into nothingness … It may be mentioned here that unlike Western medicine whose progress has been defined by a positivist value orientation built on a radical disassociation from metaphysical ideas, Indian medicine has not experienced such a split and is grounded in moral and religious interpretations of various sorts. These also govern ageing and dying. (pp. 20–21)

If life and death are therefore conceptualised as part of a life-span continuum, rather than being seen as in opposition to one another, then it is possible that old age is also socially constructed more readily in India as 'ongoing adulthood', than a conceptualisation as 'different from the adulthood that has gone before' which is more prevalent in the UK and well described by Midwinter's concept of 'post-adulthood' (Midwinter 1990). That is, if they are viewed as adult citizens regardless of their age, then citizenship obligations (incorporating 'usefulness to others') will continue to operate. If they are seen as something different from 'adults', then those obligations are less likely to be expected or valued.

Policy Relating to Eldercare in the UK and India

Adopting an approach to reciprocity in old age that is informed by meaning making, and the dialectical interplay across individual, cultural and structural forces also requires attention to the predominant eldercare policies within the countries in question. Space allows for only the briefest of overviews and I am only too aware that welfare policy in both India and the UK is currently in a state of flux as anxieties about the growing numbers of older people as a proportion of the overall population continue to cause concern on a global scale (Timonen 2008). While there are shared concerns relating to the economic implications of a rise in the number of older people who will potentially need increasing levels of social support and healthcare as they age (Bond et al. 2007; Datta 2008) other issues inform policy making at national and regional levels.

Their significance is highlighted in the following comment by Powell (2009):

> Every society in the global world are shaped by inward forces of social welfare policies for older people as well as outward forces of economic globalisation impinging on public resources and flows to people as welfare recipients: each conspiring to make welfare states uncertain in modern times. Macroscopic, global trends will be highlighted as undoubtedly powerful, yet their influence will be traced and rivalled by domestic institutional traditions in nation states. (p. 100)

Policy Informing Eldercare in the UK

In the current UK social climate there is a danger that attempts to challenge ageist ideologies and practices will be trivialised in the light of media and policy activities that focus on active and healthy ageing and arguments that, with new age cohorts entering the demographic, attitudes and expectations will change (Timonen 2008). While I applaud any move towards challenging negative stereotypes about old age, these tend to focus on the more active segment of the elderly population rather than those who are more frail and dependent and, unless that point is made very specifically, I share concerns that old age will become conceptualised in the public consciousness as less worthy of attention than it is currently because differences of experience within the elderly population will be missed. That is, media images of older people enjoying leisure activities, holidays and study opportunities may mask the isolation, marginalisation and alienation that continue to be typical of many people's later years, especially where support needs are high (Windle et al. 2011). This has the potential to dilute the anti-ageist challenge in a similar way to the situation being experienced in the women's movement, where feminists are having to respond to critics who consider the feminist movement to be redundant in modern times.

Within UK eldercare policy, reciprocity would seem to me to be implicated in a number of current debates, including those focusing on:

- Well-being (Jordan 2008); that is, whether a dependent status in old age affects what is perceived as social value. For example, initiatives aimed at facilitating the involvement of older people in local government, such as those associated with partnerships under the *Better Government for Older People* umbrella would seem to be underpinned by assumptions of competence, in that they are seen as having something of value to contribute.
- Risk management (Denney 2005; Wiseman 2011) in terms of whether initiatives or directives designed to protect dependent people from harm play a role

in denying them opportunities to lead fulfilling lives as engaged citizens, rather than alienated outsiders.
- The personalisation agenda in social care, wherein the responsibility for managing their own care needs is delegated to users of services. While such policy has the potential to help facilitate reciprocity in the lives of some individuals (including the highlighting of citizenship obligations as well as rights), it may not (according to Lymbery 2010) live up to its potential in terms of positive consequences for dependent older people. Although attracting criticism on a number of grounds (Ferguson 2007; Manthorpe and Stevens 2010) it remains high on the policy agenda at the time of writing.
- Dignity agenda—while most initiatives I have come across appear to focus predominantly on matters relating to privacy and choice, denying people labelled as dependent the opportunity to lead what they consider to be valued roles could also be seen as an affront to their dignity as multidimensional beings.
- Active ageing—the idea that people should pay more regard to, and take more responsibility for, their future health status is one that has been sanctioned not only by the UK government in their promoting of 'active citizenship', but by the World Health Organisation also (WHO 2002). Given that social engagement and support have been positively linked with well-being (Walker and Hennessy 2004; Davidson et al. 2005; Victor et al. 2009; Coote 2009; Stephens et al. 2011) it seems that the promoting of opportunities for reciprocity could usefully be associated with this agenda.

Policy Informing Eldercare in India

While there are some regional and, since devolution in the UK, national differences in policy making and welfare delivery, there is a reasonable degree of continuity in how dependent older people are conceptualised and provided for in England, Wales, Scotland and Northern Ireland.

A coherent policy across India has been more difficult to establish. On the face of it, a national policy exists in that a government initiative entitled The National Policy on Older Persons has been in existence since 1999. This policy document recognises the difficulties faced by potential family carers who either cannot fill that role because of burdensome or migratory work patterns, or who choose not to take on caring responsibilities in a modernising urban India, but promotes the view that older people should be valued members of society and not marginalised from it. Section 18 of the policy mandate is worded as follows:

> The Policy views the life cycle as a continuum, of which post 60 phase of life is an integral part. It does not view age 60 as the cut off point for beginning a life of dependency. It considers 60+ as a phase when the individual should have the choices and opportunities to led [sic] an active, creative, productive and satisfying life. An important thrust is therefore, on active and productive involvement of older persons and not just their care. (p. 3)

(www.gerontologyindia.com/national-policy.htm)

This sounds promising in terms of the value of reciprocity being sanctioned at the level of national policy, but as (Datta 2008) warns, implementation of this rhetoric at the level of individual state governance is very limited.

While the rhetoric is there, the provision of welfare support on the ground does not appear, according to Datta (2008), to be championed to any great degree at the level of national government, with the result that support services and advocacy initiatives appear to be left to non-governmental organisations and charities such as HelpAge India to provide. And, according to Kulshreshtha (2009), in addition to there being no separate government department to address the needs of older people, levels of social security are too meagre (Nair 2008) to address the poverty in which many older people live (Economic and Social Research Council briefing 2011). Add to this picture a healthcare system dominated by the private sector and within which the specialism of geriatrics is still relatively new (Ingle and Nath 2008) it would seem that those older people who need significant help from others to live their daily lives exist in a precarious and vulnerable situation unless they have supportive families, or the financial means to make lifestyle choices about where and how they live. As with the assumptions about benevolent community that underpinned 'third way' politics in the UK, and continue to underpin the ideal of 'the big society' it may be that the assumption that people will necessarily be looked after within communities is based more on rhetoric than reality.

The Phenomenological Lens

Given that my aim is to explore the affective dimension of people's experiences of reciprocity, or its absence, I have chosen a qualitative form of enquiry, as is further explained in Chap. 4.

I would argue that the research questions I propose cry out for a methodological approach that, above all, values listening. Back (2007) in discussing sociological listening, refers to the term 'thought fragments'. Alluding to the work of Arendt, in which she uses the metaphor of a pearl diver, Back makes the following comment:

> The empirical depths collected on life's surface cannot be described entirely. It is a matter of finding amid the profusion of informational debris 'thought fragments' that are the equivalent of the pearl diver's treasure. They do not illuminate the whole ocean floor, but rather they shine with histories and memories that have been transformed by the sociologist's craft (p. 21).

Such is my aim in this study—giving voice to the treasure trove of experiences and memories, the 'thought fragments' which can add to the knowledge base of the human services professions and help illuminate and explain what it is to be dependent and marginalised in old age, just as the treasure trove helps to build up a picture of the ocean floor. Moreover, it is to set these experiences and memories in the context of an empowerment paradigm: an understanding of *why* these voices should be listened to and the consequences of ignoring older people's capacity to reciprocate even when heavily dependent on others (Lustbader 1991; Boerner and Rheinhardt 2003; Jewell 2004). It has been my intention to explore, rather than to prove or disprove, and so I have chosen to set the study in the context of hermeneutics.

Key to this paradigm are the concepts of interpretation and meaning making which provide a challenge to the positivist assumption that social phenomena can be studied in an objective and value-free way. The following extract from Rorty (2007) describes the positivist assumption that *the* truth is out there waiting to be discovered:

> I shall use the term "redemptive truth" for a set of beliefs which would end, once and for all, the process of reflection on what to do with ourselves. Redemptive truth, if it existed, would not be exhausted by theories about how things interact causally. It would have to fulfil a need that religion and philosophy have attempted to satisfy. That is the need to fit everything—every thing, person, event, idea, and poem—into a single context, a context that will somehow reveal itself as natural, destined and unique. It would be the only context that would matter for purposes of changing our lives, because it would be the only one in which those lives appear as they truly are. To believe in redemptive truth is to believe that there is something that stands to human life as elementary physical particles stand to the four elements—something that is the reality behind the appearance, the one true description of what is going on. (p. 90)

Rather than working with objective facts, and searching for *the* truth, the hermeneutical mode of enquiry allows for multiple 'truths', precisely because it argues that 'facts' have to be interpreted within conceptual frameworks and through prisms of meaning. Influenced by Kierkegaard (Watts 2003) and Nietzsche (Wicks 2002) about the nature of epistemology, Gadamer (2004) introduced the concept of 'fusion of horizons'. He used it to explain how, as individuals, we move beyond our own understanding of the world (our own individual 'horizon' or perspective) by engaging with that of someone else, so that new *social* understandings and horizons form out of the interaction. By doing so, we are able to see beyond our own immediate perspective and appreciate a bigger picture.

In view of the fact that my intention is not to search for and uncover '*the* essence' or 'reality' of the phenomenon of reciprocity, but to explore differing 'realities' or horizons, I am confident that the Gadamerian perspective on epistemology validates my choosing of an hermeneutical methodology. From a phenomenological premise perception is an active process whereby people construct their own 'truths', or versions of reality, by interpreting and reinterpreting what they experience and trying to make sense of those experiences. Given that they will be processed within different contexts (different world-views for example) then perceptions will differ. From a phenomenological perspective this is not a problem, since a defining premise of phenomenology is that 'reality' does not exist independently of social life. According to Krieger (2000):

> It is not as if we can place our biology over here, and our society over there, and assume they exist independently. And we grasp this point only if we acknowledge our partial perspective. (p. 291)

As I discuss in Chap. 4, underpinning and justifying my methodology is the principle that to impose one 'reality' or assumption about how things are and should be (whether it be at the level of welfare ideology, policy making or care provision) constitutes an abuse of power because it does not validate other realities or, to use Gadamer's term, 'horizons'—that is the meaning that those in receipt of care attach to their experience of dependency. Discourse analysis alerts us to the ways in which language can be said to construct, as well as reflect, reality (Foucault 1977; Westerhof and Tulle 2007; Thompson 2011c) and so the terminology used by those I have interviewed to express their 'reality' may have something to say about the power of dominant discourses, such as medical discourse, to influence the experience of being dependent and cared for.

Conclusion

A review of the literature relating to reciprocity in two differing cultures has required an element of context setting. To have done so without exploring broader influences such as social policy, and assumptions about ageing, would be to deny the sociological dimension of people's lives, and the ways in which relationships of reciprocity acted out at a personal level are influenced by dominant ideas about what is, and what is not, appropriate for this particular cohort of people.

This study is an attempt to explore both whether dependent older people themselves value opportunities to reciprocate, and whether they are 'allowed' to by others who care for them, or who are instrumental in providing that care. So what do we already know about reciprocity in the lives of dependent older people? Bowers et al. (2011) refer to a low level of awareness of the potential for reciprocity to be life enhancing for older people in receipt of support:

> A key message so far is the low level of awareness about, and familiarity with, support based on mutual exchange and reciprocity ... a more dominant concern, however, is the prevailing perception of older people as 'passive recipients; for whom mutual support is not relevant or appropriate. (p. 5).

While research into reciprocity *per se* has a long history (Gouldner 1960; Putnam 2000), its particular relevance to the lives and care of dependent older people would seem to be relatively under-researched. It is to a review of the literature relating more directly to reciprocity in old age that I now turn my attention in Chap. 3.

References

Arber, S., & Ginn, J. (Eds.), (1995). *Connecting Gender and Ageing: A Sociological Approach*. Buckingham: Open University Press.
Arber, S., Davidson, K. & Ginn, J. (Eds.), (2003). *Gender and Ageing: Changing Roles and Relationships*. Maidenhead: Open University Press.
Back, L. (2007). *The Art of Listening*. Oxford: Berg.
Bagga, A. (2008). Gender issues in care giving, in Chatterjee et al. (2008).
Barnes, C., & Mercer, G. (2010). *Exploring Disability* (2nd ed.). Cambridge: Polity Press.
de Beauvoir, S. (1972). *The Second Sex*. Harmondsworth: Penguin.

References

de Beauvoir, S. (1977). *Old Age*. Harmondsworth: Penguin.
Bernard, M., & Scharf, T. (Eds.), (2007). *Critical Perspectives on Ageing Societies*. Bristol: The Policy Press.
Blaikie, A. & Hepworth, M. (1997) Representations of old age in painting and photography, in Jamieson et al. (1997).
Blood, I. (2010). *Older people with high support needs: how can we empower them to enjoy a better life*. Joseph Rowntree Foundation: York.
Boerner, K., & Reinhardt, J. P. (2003). Giving while in need: Support provided by disabled older adults. *Journal of Gerontology, 58B*(5), S297–S304.
Bond, J., & Corner, L. (2004). *Quality of Life and Older People*. Buckingham: Open University Press.
Bond, J. & Cabrero, G. R. (2007). Health and dependency in later life, in Bond et al. (2007).
Bond, J., Peace, S., Dittman-Kohli, F., & Westerhof, G. (Eds.), (2007). *Ageing in Society: European Perspectives on Gerontology* (3rd ed.). London: Sage.
Boneham, M. (2002) Researching ageing in different cultures, in Jamieson and Victor (2002).
Bowers, H., Mordey, M., Runnicles, D., Barker, S., Thomas, N., Wilkins, A., et al. (2011). *Not a one-way street: research into older people's experiences of support based on mutuality and reciprocity*. Joseph Rowntree Foundation: York.
Butler, J. (1990). *Gender Trouble* (2nd ed.). Abingdon: Routledge.
Butler, R. N. (1987). *Ageism, in The Encyclopedia of Aging 22–23*. New York: Springer.
Bytheway, B. (2011). *Unmasking Age: The Significance of Age for Social Research*. Bristol: The Policy Press.
Cann, P., & Dean, M. (Eds.), (2009). *Unequal Ageing: The Untold Story of Exclusion in Old Age*. Bristol: The Policy Press.
Chakraborti, R. D. (2004). *The Greying of India: Population Ageing in the Context of Asia*. New Delhi: Sage.
Chatterjee, D. P., Patnaik, P., & Chariar, V. M. (Eds.), (2008). *Discourses on Aging and Dying*. London: Sage.
Cohen, G. D. (2005). *The Mature Mind: The Positive Power of the Aging Brain*. New York: Basic Books.
Coote, A. (2009) The uneven dividend: Health and well-being in later life, in Cann and Dean (2009).
Cumming, E., & Henry, W. E. (1961). *Growing Old*. New York: Basic Books.
Datta, A. (2008) Socio-ethical issues in the existing paradigm of care for the older persons: Emerging challenges and possible responses, in Chatterjee et al. (2008).
Datta, P., Poortinga, Y. H., & Marcoen, A. (2003). Parent care by Indian and Belgian Caregivers in their roles of daughter/daughter-in-law. *Journal of Cross-Cultural Psychology, 34*, 736–749.
Davidson, K., Warren, L. & Maynard, M. (2005) Social involvement: Aspects of gender and ethnicity, in Walker (2005).
Denney, D. (2005). *Risk and Society*. London: Sage.
Dupuis-Blanchard, S., Neufeld, A., & Strang, V. R. (2009). The significance of social engagement in relocated older adults. *Qualitative Health Research, l, 19*, 1186–1195.
Economic and Social Research Council (2011) - briefing on project 'Ageing, poverty and neo-liberalism in urban south India' (www.esrc.ac.uk/impacts-and-findings/fe)
Fairhurst, E. (2003). New identities in Ageing: Perspectives on age, gender and life after work' in Arber et al. (2003).
Farndon, J. (2007). *India Booms: The Breathtaking Development and Influence of Modern India*. London: Virgin Books.
Featherstone, M., & Wernick, A. (1995). *Images of Ageing*. London: Routledge.
Ferguson, I. (2007). *Reclaiming Social Work: Challenging Neo-liberalism and Promoting Social Justice*. London: Sage.
Foucault, M. (1972). *The Archaeology of Knowledge*. New York: Pantheon.
Foucault, M. (1977). *Discipline and Punish: The Birth of the Prison*. London: Allen Lane.

Fox, N. J. (1993). *Postmodernism, Sociology and Health*. Buckingham: Open University Press.

Franklin, N. C., & Tate, C. A. (2009). Lifestyle and successful aging: an overview. *American Journal of Lifestyle Medicine, 3*(1), 6–11.

Fredricksen-Goldsen, K. I., & Muraco, A. (2010). Aging and sexual orientation: A 25 year review of the literature. *Research on Aging, 32*(3), 372–413.

Gadamer, H.-G. (2004). *Truth and Method* (3rd ed.). London: Continuum.

Giddens, A. (2009). *Sociology* (6th ed.). Cambridge: Polity Press.

Gilleard, C., & Higgs, P. (2005). *Contexts of Ageing: Class, Cohort and Community*. Cambridge: Polity Press.

Gilleard, C., & Higgs, P. (2010). Aging without agency: theorising the fourth age. *Aging and Mental Health, 14*(2), 121–128.

Gouldner, A. W. (1960). The norm of reciprocity: a preliminary statement. *American Sociological Review, 25*(2), 161–178.

Gutting, G. (Ed.), (2003). *The Cambridge Companion to Foucault* (2nd ed.). Cambridge: Cambridge University Press.

Haber, D. (2009). Gerontology: Adding an empowerment paradigm. *Journal of Applied Gerontology, 28*, 283–295.

Hafford-Letchfield, T. (2010). 'The age of opportunity? *Revisiting assumptions about the lifelong learning opportunities of older people using social care services, British Journal of Social Work, 40*, 496–512.

Havighurst, R. J., & Albrecht, R. (1953). *Old People*. London: Longman.

Heller, E. (1988). *The Importance of Nietzsche*. London: The University of Michigan.

Hill, R. D. (2005). *Positive Aging: A Guide for Mental Health Professionals and Consumers*. London: W.W Norton.

Holloway, M., & Moss, B. (2010). *Spirituality and Social Work*. Palgrave Macmillan: Basingstoke.

Hung, L. W., Kempen, G. I., & DeVries, N. K. (2010). Cross cultural comparison between academic and lay views of healthy ageing: a literature review. *Ageing and Society, 30*, 1373–1391.

Ingle, G. K., & Nath, A. (2008). Geriatric health in India: Concerns and solutions. *Indian Journal of Community Medicine, 33*(4), 214–218.

Jackson, W. A. (2009). Retirement policies and the life cycle: Current trends and future prospects. *Review of Political Economy, 21*(4), 515–536.

Jacobs, S. (2010). *Hinduism Today*. London: Continuum.

Jamieson, A., & Victor, C. R. (Eds.), (2002). *Researching Ageing and Later Life*. Buckingham: Open University Press.

Jamieson, A., Harper, S., & Victor, C. (Eds.), (1997). *Critical Approaches to Ageing and Later Life*. Buckingham: Open University Press.

Jewell, A. (Ed.), (2004). *Ageing*. Spirituality and Well-being: London, Jessica Kingsley Publishers.

Jordan, B. (2008). *Welfare and Well-being: Social Value in Public Policy*. Bristol: The Policy Press.

Krieger, N. (2000). Passionate epistemology, critical advocacy, and public health: doing our professions proud. *Critical Public Health, 10*(3), 287–294.

Kulkarni, M. N. (2009). *Admire the old and inspire the young*. January edn: Dignity Dialogue.

Kulshreshtha, S. (2009). Social security for elderly in India, www.globalaging.org/ruralageing/world/html.

Lamb, F. F., Brady, E. M., & Lohman, C. (2009). Lifelong resiliency learning: A strengths-based synergy for gerontological social work. *Journal of Gerontological Social Work, 52*, 713–728.

Laslett, P. (1996). *A Fresh Map of Life: The Emergence of the Third age* (2nd ed.). London: Weidenfeld and Nicolson.

Leibig, P. S. (2003). Old age homes and services: Old and new approaches to aged care. *Journal of Ageing and Social Policy, 15*, 2–3.

References

Lustbader, W. (1991). *Counting on Kindness: The Dilemmas of Dependency*. London: The Free Press.

Lymbery, M. (2010). A new vision for adult social care? Continuities and change in the care of older people. *Critical Social Policy, 30*(5), 5–26.

Manthorpe, J., & Stevens, M. (2010). Increasing care options in the countryside: Developing an understanding of the potential impact of personalization for social work with rural older people. *British Journal of Social Work, 40*(5), 1452–1469.

Marcoen, A., Coleman, P. G. and O'Hanlon, A. (2007) Psychological ageing, in Bond et al. (2007).

May, T., & Powell, J. L. (2008). *Situating Social Theory* (2nd ed.). Maidenhead: Open University Press.

McKee, K., Downs, M., Gilhooly, M., Gilhooly, K., Tester, S. & Wilson, F. (2005). Frailty, identity and quality of later life, in Walker (2005).

Memmers, T. M. (2004). The influence of ageism and ageist stereotypes on the elderly. *Physical and Occupational Therapy in Geriatrics, 22*(4), 11–21.

Midwinter, E. (1990). An ageing world: The equivocal response. *Ageing and Society, 10*(2), 221–228.

Minkler, M., & Estes, C. (1998). *Critical Gerontology: Perspectives from Political and Moral Economy*. Baywood: Amityville.

Moss, B., & Thompson, N. (2007). Spirituality and Equality. *Social and Public Policy Review, 1*(1), 1–12.

Moss, S. Z., & Moss, M. S. (2007). Being a man in long term care. *Journal of Ageing Studies, 21*, 43–54.

Nagendra, S. (2005). *Social Work and Social Welfare in India*. Jaipur: India, ABD Publishing.

Nair, K. R. G. (2008) A plea for a holistic approach to aging, in Chatterjee et al. (2008).

Nelson, T. D. (Ed.), (2002). *Ageism: Stereotyping and Prejudice Against Older Persons*. London: Massachusetts Institute of Technology.

Nelson, T. D. (2005). Ageism: Prejudice against our feared future self. *Journal of Social Issues, 6*(2), 207–221.

O'Hara, M. (2010). There's more to life than golf, *Education Guardian*, 12th January.

Palmore, E. (2000). Ageism in gerontological language. *Gerontologist, 40*(6), 645.

Park, N. S., Knapp, M. A., Shin, H. J., & Kinslow, K. M. (2009). Mixed methods study of social engagement in assisted living communities: Challenges and implications for serving older men. *Journal of Gerontological Social Work, 52*, 767–783.

Phillipson, C. (1982). *Capitalism and the Construction of Old Age*. London: Macmillan.

Phillipson, C. (1998). *Reconstructing Old Age: New Agendas in Social Theory and Social Practice*. London: Sage.

Pilkington, A. (2003). *Racial Disadvantage and Ethnic Diversity in Britain*. Palgrave Macmillan: Basingstoke.

Powell, J. L. (2009). Editorial: Introduction to special issue on comparative aging. *Ageing International, 34*(3), 100.

Putnam, R. D. (2000). *Bowling Alone: The Collapse and Revival of American Community*. New York: Simon and Schuster.

Rorty, R. (1989). *Contingency, Irony and Solidarity*. Cambridge: Cambridge University Press.

Rorty, R. (1998). *Truth and Progress: Philosophical Papers*. Cambridge: Cambridge University Press.

Rorty, R. (2007). *Philosophy as Cultural Politics*. Cambridge: Cambridge University Press.

Rose, H. and Bruce, E. (1995). 'Mutual care but differential esteem', in Arber and Ginn (1995).

Rouse, J. (2003). Power/Knowledge, in Gutting (2003).

Rowbotham, S. (1973). *Woman's Consciousness, Man's World*. Penguin: Harmondsworth.

Rowe, J. W., & Kahn, R. L. (1999). *Successful Aging*. New York: NY, Dell.

Rubenstein, D. (2001). *Culture, Structure and Agency; Towards a Truly Multidimensional Society*, London:Sage.

Shachter-Shalomi, Z., & Miller, R. S. (1997). *From Age-ing to Sage-ing: A Profound New Vision of Growing Older*. New York: Grand Central Publishing.

Sibeon, R. (2004). *Rethinking Social Theory*. London: Sage.

Srikrishna, L. (2006). Old age homes preferred by many elders, *The Hindu*, 25th November.

Srinivas, S. & Vijayalakshmi, B. (2001). Abuse and neglect of elderly in families, *The Indian Journal of Social Work, 62*(3) 464–479.

Stephens, C., Alpass, F., Towers, A., & Stephenson, B. (2011). The effects of types of social networks, perceived social support and loneliness on the health of older people: Accounting for the social context. *Journal of Aging Health, 23*, 887–911.

Stones, R. (2005). *Structuration Theory*. Palgrave Macmillan: Basingstoke.

Thompson, N. (1993). *Anti-Discriminatory Practice*. Basingstoke: Macmillan.

Thompson, N. (2010). *Theorizing Social Work Practice*. Palgrave Macmillan: Basingstoke.

Thompson, N. (2011a). *Promoting Equality: Working with Diversity and Difference* (3rd ed.). Palgrave Macmillan: Basingstoke.

Thompson, N. (2011b). *Effective Communication: A Guide for the People Professions* (2nd ed.). Palgrave Macmillan: Basingstoke.

Thompson, S. (2007). Spirituality and Old Age. *Illness, Crisis and Loss, 15*(2), 169–181.

Timonen, V. (2008). *Ageing Societies: A Comparative Introduction*. Maidenhead: Open University Press.

Tokarski, W. (2004). Sport of the Elderly. *Kinesiology, 36*(1), 98–104.

Townsend, P. (2007). 'Using human rights to defeat agesism: Dealing with policy-induced structured dependency', in Bernard and Scharf (2007).

Tulle-Winton, E. (2000). Old bodies. In P. Hancock, B. Hughes, K. Jagger, R. Paterson, E. Russell, E. Tulle-Winton et al. (Eds.), *The body, culture and society: An introduction*. Buckingham: Open University Press (2000).

Twigg, J. (2004). The body, gender and age: Feminist insights in social gerontology. *Journal of Ageing Studies, 18*, 59–74.

Varma, P. K. (2005). *Being Indian: The Truth about Why the Twenty-first Century Will be India's*. London: Penguin.

Victor, C., Scambler, S., & Bond, J. (2009). *The Social World of Older People: Understanding Loneliness and Social Isolation in Later Life*. Maidenhead: Open University Press.

Walker, A. (1981). Towards a political economy of old age. *Ageing and Society, 1*, 73–94.

Walker, A., & Hagan Hennessy, C. (Eds.), (2004). *Growing Older: Quality of Life in Old Age*. Maidenhead: Open University Press.

Watts, M. (2003). *Kierkegaard*. Oxford: Oneworld.

Westerhof, G. J. and Tulle, E. (2007) 'Meanings of ageing and old age: discursive contexts, social attitudes and personal identities', in Bond et al. (2007).

Whitbourne, S. K. & Sneed, J. R. (2002). The paradox of well-being, identity processes and stereotype threat: Ageism and its potential relationship to the self in later life', in Nelson (2002).

Wicks, R. (2002). *Nietzsche*. Oxford: Oneworld.

Wilson, G. (2000). *Understanding Old Age: Critical and Global Perspectives*. London: Sage.

Windle, K., Francis, J., & Coomber, C. (2011). *Preventing loneliness and social isolation: Interventions and outcomes, SCIE Research Briefing 39*. London: SCIE.

Wiseman, D. (2011) A "four nations" perspective on rights, responsibilities, risk and regulation in adult social care, York, Joseph Rowntree Foundation.

World Health Organization. (2002). *Active Ageing: A Policy Framework*. Geneva: WHO.

Chapter 3
Reciprocity and Old Age

Introduction and Background

Having drawn on existing literature to provide context and justification for this research, I move in Chap. 3 to situate it in the literature base that focuses more directly on reciprocity itself. I am aware that there is an extensive and long-standing body of literature around reciprocity, much of which has a foundation in economics, philosophy, psychology, politics and sociology, and I draw on some of that literature below. In several ancient philosophical traditions reference can be found to what is often referred to as 'the golden rule'—an unwritten but guiding principle that has underpinned the world views of many religions, including Christianity and Hinduism, and which enshrines the expectation that we should behave towards others as we would expect them to behave towards us. As Warren (2008) comments:

> Reciprocity is the basic norm of social exchange–so basic it is built into most ethical and cultural systems, through one or another formulation of the Golden Rule. If I do something for you, I then expect to be able to call on you in a time of need at some point in the future. (p. 139)

In what is considered to be a seminal work, Gouldner (1960) refers to reciprocity as a shared 'moral norm', theorising the concept in terms of power differentials that have the potential to lead to inequality and social disruption unless that moral norm is observed. Uehara (1995) does not deny the 'moral norm' argument but maintains that there is not a great deal of evidence to suggest that balanced social support relationships exist. Rather, she suggests, individuals tend to prefer to 'underbenefit' from reciprocal arrangements of social support–that is, they tend not to expect a similar level of support to that which they themselves have provided for others. These understandings fit with my own use of reciprocity to describe a sense of 'usefulness' derived from being able to 'give back' in general terms, rather than a tit-for-tat obligation.

Explanations based on cost-benefit analyses assume rationality and I am more convinced by arguments based on the power of ideas to influence thoughts and actions, such as in the works of Gramsci (Nowell-Smith 1998) and Foucault (1977), than I am in the suggestion that reciprocity operates in a rational way, untouched by social expectations about, for example, the 'appropriateness' of aspirations to grow intellectually and spiritually and to feel 'useful' in old age.

Narotzky and Moreno (2002) make the point that, while the concept of reciprocity has remained relatively uncontested since being conceptualised as a moral norm by Gouldner, and having a focus as it does on the maintenance of social stability, it takes insufficient account of what they refer to as opposing positive (described as shared morality) and negative (a suspension of, or change to, the moral order) aspects which they argue contribute to its complexity. This extract from their work is significant in relation to the disenfranchisement of older people and the alienation they can experience, because it challenges the idea that there is a single, shared and unchanging moral norm:

> Reality is complex and involves different but connected social systems undergoing constant transformation, with different groups and subjects being mutually produced as subjects and as groups. These social groups will produce moral codes or systems of mutual obligation and responsibility that will exert pressure to modify or save particular social configurations and systems of inequality. One can easily appreciate how both aspects of reciprocity are constantly and simultaneously present as the tension between beneficial or predatory outcomes of social relations…We believe that the concept of reciprocity still has a useful role to play in analysing social processes providing it incorporates negative reciprocity. The emphasis on equality and balanced exchange as the starting point for a reciprocal relation has only served to hide the imbalance and ambiguity in reciprocal relations and their capacity to generate, reproduce and transform systems of inequality in reference to a field of moral forces where conflict and ambivalence prevail. (p. 301)

Given that the literature base on reciprocity in general is so vast, and that space is limited, I choose not to present anything more in the way of a general overview, although I draw on some aspects of the wider literature where I feel it can help to illuminate a particular point.

In order to keep this study phenomenologically grounded, I use the key concepts flagged up in the introductory chapter to inform the structure of latter part of the review. That is, I highlight literature relevant to the inter-relationships between reciprocity and the following:

- social space;
- social time;
- meaning making at the person/spiritual level; and
- meaning making at the level of discourse and institutionalised patterns of power.

The significance of these concepts will be explored briefly in the introductions to their relevant sections below but, before moving on, I refer back to my choice of the word 'highlight' in the preceding sentence. Although I make the point in this chapter that literature relating to reciprocity as experienced specifically by older people in receipt of formal care is limited, when explored in relation to the concepts listed above, a much larger literature base opens up as inter-relating issues emerge–any one of which could provide the basis for a paper on the significance of its relationship with reciprocity. In light of this, I make it explicit here that my intention is only to flag up the existence of literature that sets a context for my research, and helps to justify it as a development of the existing knowledge base.

At times, the points may seem unrelated and underexplored but I make no claims to comprehensiveness or in-depth analysis, my aim being only to reflect ways in which I feel the themes I am exploring have already been addressed in ways that feed into an understanding of the significance of reciprocity in old age. While lack of space does not allow me the luxury of critique or analysis here, the issues thrown up by this review may provide fertile ground for further research and critical analysis by myself or others in the future.

I have already made reference to the literature base which *specifically* addresses reciprocity as experienced by older people in receipt of formal care, and which *specifically* pays attention to phenomenological concerns and methodology as being fairly minimal. Some of the literature I draw on is therefore included in this review more as a justification for why I have been moved to choose the research focus that I have, than as a reflection of extant literature that seeks to explain its existence. To highlight what we already know about the subject I structure this discussion by drawing on three concepts that, to my mind, encompass the field with which I am concerned:

- *the psychosocial impact*: that is, I refer to literature which reflects that reciprocity is not experienced atomistically in a social vacuum, as if the context in which reciprocal relationships take place is not significant.
- *social capital and relationships*: this reflects the significance of relationships more specifically. I use the term social capital here to refer to the social contacts that people can draw on as 'capital' in order to further their own interests or help them in times of need. Coleman (1988); Putnam (2000) and Lin (2008) have been influential in developing theory in this area, as too has Bourdieu (1986), whose use of the term resonates with my own research because of his focus on how operating in the 'right' environments has the potential to promote or deny opportunities for social support.
- *well-being*: the senses (be the physical, emotional, psychological or spiritual) in which we feel good about ourselves, and the ways in which reciprocity or its absence can leave people feeling enriched or depleted, valued or otherwise.

These concepts interrelate, and so the discussions within them may overlap or have aspects that could justify their allocation to anyone, or all three, of the categories. I use this framework only as an aid to exploring otherwise seemingly disparate literature, rather than proposing them as discrete categories.

Social Space

In its broadest sense, social space refers to the space in which we live out our lives–the space in which relationships operate. As Giddens (2009) suggests, 'all interaction is situated' (p. 147).

In the sense that it can be said to be situated in space, there are two aspects which can be seen to have significance for this study of reciprocity:

(1) the concrete physical environments in which social interactions take place—home, institutions and so on.
(2) space in the abstract sense—the being out there in the world and interacting with others.

Both aspects feature in this very brief account of existing literature that theorises, or has significance for, reciprocity.

The Psychosocial Impact of Social Space

Of the many relevant and important aspects I could have explored here, I highlight the following five as particularly significant:

(1) The physical space in which older people live out their lives is significant in terms of reciprocity because of the different cultures that operate in different environments. It is likely that those dependent older people who continue to live in their own homes are more able to 'live by their own rules' than those living in care homes where there is often a tension between managing the interests of the residents and those of the institution (Goffman 1961; Renzenbrink 2004; Peace et al. 2006). But even those people in receipt of domiciliary care support often have their preferences subordinated to the needs of busy care workers and the operational needs of care agencies (Commission for Social Care Inspection 2006; Social Care Institute for Excellence 2009). Regimes leave little room for spontaneity (Thompson 2002) as do living arrangements with little in the way of facilities or privacy (Renzenbrink 2004; Hall et al. 2009). These factors have the potential to chip away at an older person's identity as someone competent enough to decide for themselves what do, how and when to do it and to what end.
(2) Peace et al. (2006) and Ryvicker (2009) echo this in suggesting that a sense of place is also a sense of person—the space in which we live playing a part in the construction of our identity. Especially for those living with other dependent residents in residential settings, it can be difficult for 'home' not to be defined in terms of dependency. And, furthermore, for that to feed into how other aspects of identity (including that relating to the ways in which one feels proud of skills or knowledge that one can share), become subsumed under a blanket categorisation as being 'in need'—a label which is a pre-requisite for admission to care facilities or the provision of domiciliary-based care. MacKinlay (2001) cites Tournier (1972) as alluding to the need for people to conceptualise themselves not as 'human doers' but 'human beings' when they reach old age, if they are to be able to make sense of the new identities being accorded them, but I would argue that there is no reason why that acquiescence needs to take place if the will to facilitate human 'doing' exists in those settings.

(3) If we accept the argument that social life involves intersubjectivity (Crossley 1996; Gergen 2009), so that we need interaction with other people in order to make sense of our place in the world, then anything that contributes to their isolation from the communities in which they have lived their lives previously has the potential to deny dependent older people opportunities to be 'useful' members of those communities. Where opportunities for older people to continue to make a contribution to those communities are not facilitated, the potential for them to feel that:

(a) they are a burden on others; and
(b) they have no direction or purpose in life,

may therefore be enhanced. And where they do not receive social validation as contributors, this may contribute to, and reinforce, a sense of alienation they may be feeling and their spiritual well-being compromised as a consequence. That is, a sense of ontological insecurity related to a perceived mismatch between the person they consider themselves to be, and the person others perceive them to be. This sense of unease about one's own identity is explicit in Fromm's understanding of alienation: 'By alienation is meant a mode of experience in which the person experiences himself [sic] as an alien. He has become, one might say, estranged from himself.' (Fromm 1955, p. 110)

And so, removing them from communities characterised by differences of interest and ability to ones populated only by other dependent people–perceived within a 'care model' to be undifferentiated by differences of interest and ability (Thompson and Thompson 2001)–reduces the space in which a sense of 'usefulness' through contribution can be enacted, unless existing and new connections with wider communities are facilitated.

(4) This isolation can be compounded by broader policies such as public transport and town planning, which, where they fail to address:

(a) difficulties experienced by older people with mobility problems (Manthorpe et al. 2006);
(b) the particular difficulties of older people travelling in and between rural areas (Peace et al. 2007) and
(c) personal safety on streets and public spaces (Holland et al. 2005)

compromise the opportunities for engaging with others that are contributory factors to the construction and maintenance of self-identity and self-esteem.

(5) Current literature also highlights another aspect of social space that has significance in relation to reciprocity–that of 'virtual' social space. The rise of the Internet has opened up new communities, and ways of interacting with others within those communities, so that geography becomes less of an issue in terms of communicating and engaging with others (Hiller and Franz 2004). Consequently, physical frailty or incapacity which results in isolation may be less of a barrier to engaging with, and finding ways to 'give back to', wider

communities than it once might have been (Bradley and Popper 2003; Galit 2010). As Maya-Jariego and Armitage 2007 suggest:

> Modern communications have weakened the traditional relationship between physical setting and social space, enabling participation in multiple communities simultaneously. Physical space is no longer necessary or a guarantee for participation (p. 743).

However, if the potential exists for the opening up of this form of social space as one in which older people can play a 'useful' role by, for example, offering advice or mutual support, or by sharing information or sources of information, then that potential can only be realised where the will to facilitate it is also there. Though Whittall and Grace (2002) report on an initiative to enhance communication between nursing home residents and schoolchildren, to date I can find little evidence that older people dependent on formal care, especially in institutions, are routinely being provided with the opportunity to engage with others in virtual communities through the medium of Internet technology.

Social Capital, Relationships and Social Space

There are three important aspects to consider here:

(1) Although social capital has been criticised on the basis that it has become too broad a concept to be useful, I would agree with Glanville and Bienenstock (2009) that it is its breadth that makes it useful as an integrating concept. From my perspective it is helpful because this study of reciprocity relates to the nature of, and opportunities for, 'valued' social interaction. Since social capital relates to the number and quality of relationships with other people, then I would argue that it has some validity, especially as it concerns those on whom we can rely for support. It would seem fair to say that reciprocity both depends on, and contributes to, our reserves of social capital. That is, if our reserves of social capital are low, then opportunities for reciprocity between people in those networks are low. Conversely, if our reserves of social capital are high, then opportunities for reciprocity are also high.

Where older people are dependent to the extent that they are unable to easily leave the places in which they live, and become reliant on others to visit them, then there already exists an imbalance in the relationship which might colour how they are perceived–in this case as someone who is defined as 'in need' (of having items brought in for them, favours or procedures done for or to them, for example)–even though this is often a misrepresentation and could be reframed by a change of perspective. In her 1991 publication, Lustbader makes that very point in offering a vignette from which I present the following excerpt:

> [he] goes on to tell the story of discovering a common passion for genealogy with one of his clients: "So we welcomed each other's craziness and became real friends at once". Suddenly the homebound man was not receiving charity but rather was participating in a pleasure. He was not "simply being visited," but rather was granting his visitor a chance to indulge and amplify his own interests. (p. 28)

This is supported by Andrews et al. (2003), who also describe how formally arranged visits to housebound people by volunteers designated as 'befrienders' can become relationships of friendship if elements of reciprocity develop. Social capital has been described in terms of 'bonding' and 'bridging' potential–the former describing mutually reinforcing networks between people who are socially alike and the latter describing networks between people who are socially different (Putnam 2000). I would suggest that the 'horizontal ties' and 'vertical ties' referred to by Fennema and Tillie (2008) better account for the power imbalances that inform the promoting, or otherwise, of opportunities for engaging with others in ways that have the potential to make dependent older people feel valued. Horizontal ties, or connections, refer to relations between people who are on a similar footing in terms of power. Vertical ties relate more closely to relations where there is a power differential. Given the social and physical space in which older people dependent on formal care live out their lives–either in their own homes or in institutions–then it can be assumed that most relationships they experience are characterised by an imbalance of power because of their vulnerable position in the carer/cared-for dynamic.

(2) In this sense, the significance of social space becomes apparent at an institutional level as well as a personal one. At a personal level, if someone has few or no visitors and cannot sustain meaningful relationships by getting out to attend societies, clubs and so on, then opportunities for reciprocity–to feel engaged and 'useful'–become reduced. At an institutional level, if the only or main relationships people have are with staff, then opportunities for reciprocity may not only be reduced in number, but also subject to the attitudes and practices of staff who themselves may be influenced by dominant discourses to see the people in their care as unidimensional beings, and the institution as the only space in which they can operate. This is something that Bowers et al. (2009) highlight from their research:

> Whilst the range of mutual support options is wide, the research has also found that perceptions of, and attitudes towards, older people with high support needs is narrow, as indicated above by the attitudes of some of the professionals involved or contributing to the study to date. Older people with high support needs are still largely perceived as people 'in need of support' who need to be 'taken care of'—rather than as citizens with rights, responsibilities and contributions to make. (p. 5)

(3) The 'commodification of care' thesis may also throw some light on the significance of social space–in the sense of personal space being the space in which meaningful relationships take place. This term is used by Ungerson (2000) to describe how care has increasingly become something that attracts a price–a commodity that can be paid for, and therefore undertaken not necessarily out of love or duty, but for wages. As I am researching people's understanding of their experiences of reciprocity, or its neglect, the consequences of this trend may be significant for the meanings that people attach to their experience. That is, in a social space where care is provided informally by family or friends, and without payment, the obligation to reciprocate may be felt more strongly than in a social space where someone is

receiving support that they are either themselves buying as a commodity or is being bought for them by family or the state. On the other hand, paying for care may help to maintain a degree of control so that those receiving it feel less dependent (Garey et al. 2002).

Well-Being and Social Space

And finally in this section, there are three further points to consider in connection with spiritual well-being:

(1) Well-being is a broad concept which incorporates a number of dimensions, including physical, psychological and spiritual aspects in which all have significance for quality of life (Walker 2005; Jordan 2008; Victor et al. 2009; Forte 2009). For those dependent on formal care, all of these aspects are important in that the quality of care relationships has implications, not only for an individual's physical health, but also for their sense of identity and self-worth. This is highlighted in evidence from the perspective of nursing home residents (Bowers et al. 2001) many of whom identified 'care-as-relating' as an aspect of care provision that they perceived as life enhancing:

> Care-as-relating residents spoke less about the technical aspects of care (the how and when described by care-as-service residents), but more about the signs of individualized affection and friendship they found in the care they received... care-as-relating residents saw reciprocity as evidence of good relationships, and thus of good quality care. Residents often discussed reciprocity in terms of sharing invisible or past identities. An aide would share with the resident previously unknown personal details related to her life outside work; in turn, the resident could share personal identities from his or her past. "Good" aides were described as attending to those identities as they provided care. By doing so, these aides were acknowledging resident selves other than those related to old age, illness and disability. As one resident noted, a good aide was one who could "see me as not just an old lady or someone with bad knees and a catheter to clean". (p. 542)

(2) While the focus might not be strictly on reciprocity, Johnson (1998) and Knight and Mellor (2007) challenge the provision of 'one-activity-fits-all' sessions both in institutional settings and daycare facilities which, although they may be underpinned by a desire to encourage physical and mental agility, often fail to see beyond the stereotype to the individual's unique personality, history and aspirations–in essence their spirituality or authentic selves. Clow and Aitchinson (2009) suggest that the once-common (and, though I can only comment from my own observations and anecdotal evidence, in my opinion *still* common) programmes of bingo and sing-along have, in many cases, been updated to include what they consider to be more innovative and challenging activities. However, I would suggest that such sessions seem still to be premised on a goal of 'staying active and engaged' rather than on that of 'staying *me*', an endeavour which requires attention to facilitating opportunities to continue with *existing* activities and relationships that have defined who a

person considers themselves to have been, and wants to continue to be–reciprocity being part of that.

In support of this premise, one of the key messages to emerge from the service user feedback aspect of a recent study of initiatives to address loneliness and social isolation in old age (Windle et al. 2011) is that enjoyment of group activities would be enhanced if organisers took more account of user preferences.

(3) This neglect of a spiritual need for biographical continuity echoes the sense in which Midwinter (1990) describes a tendency for dependent older people to be conceptualised as 'other'–someone different and 'less than' their earlier selves. Where this happens, the meaningful ways (meaningful in the sense that they relate to a life and relationships prior to becoming dependent on others) in which people could be helped to feel 'useful' within the new social space they have come to inhabit can be overlooked.

This recognition of the well-being that can arise from feeling 'useful' is evident in advice emanating from a care staff bulletin which focuses on helping people to manage the transition from independent space to supported or residential living spaces (Help the Aged 2007)–that is, to consider giving over 'responsibility for something', a greenhouse, for example–in situations where an individual could gain self-esteem from giving back his or her gift of, in this case, gardening expertise. In her evaluation of a communal gardening project at a nursing home, Raske (2010) also emphasises the positive link between *meaningful* daily activities and quality of life.

In evaluating the 24 research projects relating to quality of life [QoL] that were carried out under the umbrella of the Growing Old Programme [GO], Walker and Hagan Hennessy (2004) make the following point:

Taken as a whole, the findings from the GO programme underscore the nature of QoL, in older age as inherently multidimensional, dynamic, actively constructed with reference to past and present selves, and context dependent. (p. 13)

This supports the premise that social space, both in terms of physical environment and the arenas in which relationships take place, is a significant factor in relation to the maintenance of a valued identity. Where social space does not allow for the social participation and engagement that are assumed to be crucial to well-being (Walker and Hennessy 2004; Davidson et al. 2005; Victor et al. 2009; Coote 2009) or does not allow it to be experienced on equal terms such that the older person's contribution is valued, the implications for spiritual well-being are clear because one's spiritual dimension can be said to incorporate self-worth and self-esteem. Studies have already indicated the significance of reciprocity for a sense of well-being in old age (Moriarty and Butt 2004). However, in light of, and perhaps to some extent due to, the persistence of a medical model of care, with its strong emphasis on deficit in old age (Palmore 2000; Means 2007), the association between reciprocity and spiritual well-being appears to remain underresearched.

I have drawn on a broad range of literature to support my premise that social space–both in terms of the environments in which people live, and the space in which relationships operate–has significance for an exploration of the meanings people attach to reciprocity. Where aspects of social space limit opportunities for reciprocity, or foster cultures where reciprocity is not seen as relevant, older people in receipt of formal care can find themselves having to re-evaluate their sense of identity as competent people with something to offer others in return for what they take from life. To ignore what, for many, amounts to an existential crisis (Moss 2005; Tanner 2010) is to seriously underestimate the significance of reciprocity for the spiritual well-being of older people in the circumstances of this study.

Social Time

While social life is lived out in a spatial dimension, it is also set within a temporal context. That is, it is also experienced in a state of flux, in that situations, relationships and biographies do not remain static (Sartre 1963; Tanner 2010; Bytheway 2011). In the sense that social life is dynamic, interactions and their consequences are constrained by structures and attitudes which are also constantly changing, so that we no more live in a temporal vacuum than we do in a social or moral one.

So, what do we already know about the significance of social time for older people in receipt of formal care?

The Psychosocial Impact of Social Time

Much of what I have already referred to in terms of social space also has a temporal element. As this seems self-evident, I will not revisit those issues here but, instead, focus on additional literature which adds to an understanding of the sense in which social life can be seen as a continuum.

(1) There is a good deal of evidence to suggest that attention is paid in social care provision to the interaction between past and present, as in life history, biographical, reminiscence and life history approaches (Gulette 2004; Gibson 2011; Bornat and Tetley 2011), but it is less clear how much the future dimension of hopes and aspirations is taken account of.
(2) In discussing the concept of personhood, Thompson (1998) highlights the importance of recognising future aspirations when he draws on what Sartre (1963) refers to as the 'regressive-progressive method' for an analysis of social life.

> This method was presented by Sartre as operating at two levels, macro and micro. The macro level has much in common with Marxism in terms of human praxis as the motor force of history… The micro level represents the process of self-creation, the interplay between past experience and future aspirations that guides both our actions and others' perceptions of us. Who we are depends on a complex set of interactions between who we are and who we want to be. (p. 699)

As discussed earlier, the interplay between past, present and future is also evident in what Heidegger (1962) referred to as 'thrownness'. That is, we are 'thrown into' or find ourselves in, a present that has been led up to and shaped by experiences in the past, but is also shaped by the future, in connection with our vision of who we want to become. It could be argued that, if reciprocity has been a life-affirming element of the past, then it is likely to be part of people's vision–their existential project–for the future. But for that project to be enacted, opportunities for reciprocity have to exist. If the context of people's lives is such that those projects are not recognised or validated because they are not seen as having a future element to their lives, then opportunities for connectedness–the basis on which reciprocity operates– are unlikely to be facilitated.

(3) This sense of continuing identity fits well with the proposition that older people engage in processes directed towards 'sustaining the self' (Tanner 2010)– that is, the construction of a psychobiography which incorporates a sense of who they are and what is significant about their particular life journey. It is also incorporated in the concept of 'habitus' (Bourdieu 1984) which refers to an individual's own 'culture'–the set of behaviour patterns, emotional characteristics and so on, which make them unique as a person. But, given that individuals live in a social world (Crossley 2005; Gergen 2009; Giddens 2009), neither psychobiography nor habitus will have been constructed randomly, but influenced by the broader cultures in which people have lived their lives. As such, it can be seen to be more than simply fixed traits of personality. What distinguishes habitus from habit, according to Garrett (2010), is the sense in which it is has a 'generative capacity' (p. 1755). That is, it is within the power of individuals to reflect on, and adapt, their sense of who they are and how they want to live their lives. And so it can be argued that identities (such as self-identification as someone who has the capacity to be a useful member of society) are not constructed within a vacuum, unmediated by changing prevailing ideologies or discourses about what is age appropriate.

(4) Furthermore, if the dependent older person's project is to stay being the person they have always considered themselves to be (in this case, someone who can give as well as receive), for that project to be facilitated it matters whether others share that sense of a continuing biography. However, two concepts strongly associated with a bio-medical perspective have the potential to provide a challenge to the conceptualisation of old age as a continuation of a 'life in progress' by suggesting that it can be conceptualised as being the end of one stage (the multidimensional person) and the beginning of another (the unidimensional person):

   (a) Studies of adaptation to chronic illness often refer to 'biographical disruption', a term attributed to Bury (1982) which refers to health crises causing a disruption to the usual way in which one lives one's life, and of the self-identity one has already constructed by that point.

   A fairly recent critique by Godfrey et al. (2004) highlights the extent to which the meanings that people attach to their experiences of sudden or chronic illness are significant for how they integrate any consequences into their lives, and questions the assumption that dependent older people necessarily experience 'biographical disruption' when they become dependent on others. Despite the fact that chronic impairment does not negate the potential for people to give while also being in need (Boerner and Rheinhardt, 2003), it would seem that the idea that older people leave their former 'competent' lives behind when they become dependent on formal care continues to be influential (Bowers et al. 2011).

> (b) The 'body drop' thesis also gives credence to the idea that old age is necessarily characterised by a decline in health. The phenomenon of 'body drop' is explained by McKee (1998) in terms of a terminal decline into dependency following falls and similar events, which, having been attributed to old age, contribute to older people's self-perception as old and in irreversible decline, such that their resilience is adversely affected. As a consequence, they may come to assume that there little point in looking to personal growth, continuity of self-image as a valued person and so on. However, the thesis of irreversible decline has its critics. For example, Cohen (2005) refers to evidence which contradicts the inevitability of decline in old age by arguing that the brain does not necessarily deteriorate with advancing age, but has the capacity to grow and adapt. It seems from Cohen's evidence that some areas are capable of being activated or re-activated by experience and therefore are not seen to function at maximum capacity until the later stages of life. A challenge to the assumption of inevitable decline is also supported in Indian gerontological literature (Naratajan 2000).

(5) For people to have opportunities to feel 'useful' (be it in ways that they always have done, or in different ways) and where they need help in facilitating this, other people's values come into play. I refer not only to the personal values of individual eldercare practitioners, and those underpinning the organisational cultures within which they work (as discussed earlier with reference to the research carried out by Bowers et al., 2011), but also the prevailing political ideologies that inform those organisational cultures.

As those political ideologies change, so do policy directives, as is evident from policy changes in the management of risk in old age (Bornat and Bytheway, 2010), which have an impact on the extent to which people are 'allowed' to operate in ways, or environments, where they have formerly felt competent to reciprocate.

Social Capital, Relationships and Social Time

The following points are significant in respect of the above:
(1) I have talked about the importance of social contacts in people's lives if they are to be experienced as fulfilling, but it needs to be borne in mind that

relationships themselves are not static. Those dependent on formal care often experience a loss of social contacts, be it through relocation or being unable to easily leave one's home (Renzenbrink 2004; Victor et al. 2009). In that sense, then, the elements of reciprocity that may have characterised those relationships are also likely to be lost in the process. For many, then, the crises that lead to a dependency on formal care can force a re-evaluation of one's life in the wake of existing social networks being disrupted or dismantled.

The meaning reconstruction approach, as evident in thanatological literature (Neimeyer and Anderson 2002; Neimeyer and Sands 2011) is helpful for understanding how loss challenges the taken-for-grantedness of our lives, and undermines the sense of security derived from having made at least some sort of sense of those lives, and our roles and relationships within the worlds we inhabit. It also addresses the significance of time for enabling losses to be reframed in the context of new situations. In Thompson (2002) Neimeyer and Anderson make the following point:

> When traumatic loss confronts us with a profound disruption of our 'assumptive world' (Janoff-Bulman and Berg 1998), that sustaining but largely unspoken network of beliefs in a predictable life, a benign universe and a worthy and resourceful self, we are faced with the onerous task of revising these taken-for-granted meanings to be adequate to the changed world we now occupy. Simultaneously, we must deal with urgent questions about what this loss signifies, whether something of value might be salvaged from the rubble of the framework that once sheltered us, and who we are now in light of the loss or losses sustained. All of this questioning plays out on levels that are practical, existential and spiritual, and all of it is negotiated using a fund of meanings (partially) shared with others, making it as much a social as a personal process. (p. 48).

From a meaning reconstruction perspective, then, it can be argued that it is not only the social contacts referred to above that are likely to be lost due to relocation or enforced isolation because of increasing dependency on others, but the *meaning* attached to those social contacts. For many people, the loss of a person or community is also the loss of an opportunity to feel of value to that person or community, and may contribute to the sense of alienation described by Fromm (1955), as discussed earlier. However, if that sense of alienation is to be addressed, the experience of losing social contacts (and the social capital that can be accrued from having them) has to be recognised as a loss in the first place. As Lloyd (2002) suggests, this cannot be taken for granted. At a time of life when the potential for social contacts to be lost through death or relocation is high, it is therefore important that a broad conceptualisation of loss informs our understanding of the significance of reciprocity for spiritual well-being.

(2) One of the opportunities that can be lost when social interaction is limited is that of an intergenerational sharing of knowledge and support. Research evidence (Hatton-Yeo 2006) indicates a rise in the numbers of initiatives where older people are involved in projects where they support the learning of children and young adults–helping with literacy, informing their understanding of history through sharing of their life stories and so on. It is difficult to know whether the learning is bi-directional or not, but such initiatives can still be seen as evidence of older people having an opportunity to give something

back, constituting a form of 'bridging' social capital (Gray 2009) that can be to their interest. It would be interesting to know the extent to which those who are dependent on formal care provision are invited to be part of such initiatives–especially where they live in residential settings, or are largely hidden from view in their own homes–and the extent to which the current emphasis in the UK on protecting vulnerable people from harm can be said to be impinging on people's rights to take part in reciprocal exchanges.

(3) Intergenerational initiatives have the potential to promote old age as just one more stage along a lifelong learning continuum, rather than one no longer associated with intellectual or spiritual growth (Groombridge 1982; Stanton and Tench 2003; Hatton-Yeo 2006). They fit well with the concept of 'sage-ing', whereby old age is reframed as a time of expertise and wisdom sharing, rather than one characterised by withdrawal into self-interest (Shachter-Shalomi and Miller 1997). Their 'sage-ing' thesis proposes that institutions housing older people could potentially become centres of learning–repositories for the wisdom garnered from long and full lives–rather than being centres where leisure pursuits fill otherwise empty days. This is perhaps an idealistic premise but, as part of a movement to reframe old age as a time of learning and involvement in education (Hafford-Letchfield 2010), it is an interesting addition to literature which challenges the association of old age and decline.

(4) It does not, however, address the issue of whether what is perceived as 'the gift of wisdom' in one age cohort is received as such by younger ones, given that the value of skills and knowledge are context dependent (Jones and Miller 2001). For example, on first inspection, an intergenerational initiative between students and older people—'*Eldertainment: generating friendships between young and old*' (www.eldertainment.co.uk)—sounded promising to me in terms of a good example of what older people could offer young students in the way of experience and expertise. Interestingly in terms of epistemology, however, the differential value accorded each cohort was reflected in the fact that the initiative required the students, but not their potential older 'partners', to submit a Curriculum Vitae.

Well-Being and Social Time

Several issues relating to temporality and change are pertinent here, especially in relation to the spiritual well-being associated with opportunities for reciprocity:

(1) Chatterjee (2008) refers to Indian tradition conceptualising community as the smallest significant social unit, which may have had implications for perceived obligations in terms of reciprocity. This may be changing in urban India where 'the cult of the individual' seems to be gaining prominence (ibid) but, where an obligation to community remains strong, taking someone out

of that community, because of frailty, has the potential to also remove opportunities for contributing to it, thereby eroding that person's understanding of themselves as valued.

(2) This relates to the concept of world-view, which incorporates how one sees the world and one's place within it (Moss 2005). It is not necessarily the case that being able to give as well as receive is incorporated into every person's concept of 'a good life'. And so, while I argue here for reciprocity in old age to be better understood and facilitated, I recognise that it should remain a matter of choice, and therefore not forced on people who do not want it. Indeed Godfrey et al. (2004) comment that some older people may take the view that, having given of themselves earlier in their lives, they may be happy to take a back seat and accept a more passive role when they find themselves in a situation of dependency. For many, though, reciprocity *is* important for well-being and, in situations where opportunities are limited, it may be necessary for some to reframe experiences (as discussed below) so that they fit with an existing world-view where they have a valued place, rather than have to accept that things have changed.

This may help to explain the use of 'deference as reciprocity', a concept highlighted by Beel-Bates et al. (2007). For example, someone may agree to go into a nursing home, or accept care support, not from personal choice, but in order not to be a burden on another person or persons. This framing of the decision as a gift to those who might otherwise have been inconvenienced or burdened may then allow the person entering into their new relationship of dependency on others to retain a sense that they still have something to give back. This may be akin to the 'mastery' described by Chokkanathan (2009) whereby older Indians can be seen to defer to family members who care for them, and ultimately to their belief in karma. By doing so, they can be seen to maintain some sense of control by not getting in the way of their destiny unfolding.

(3) One further point that highlights the links between reciprocity, social time and well-being is the extent to which changing circumstances can be seen to have the potential to be transformational–opening new and creative ways to enhance personal well-being through engaging with the world (Bianchi 2005). If a person's 'habitus' has incorporated reciprocity then changing circumstances which do not allow for it, or diminish its value, can cause a sense of dissonance that has implications for well-being. However, as we have seen (Garrett 2010), one's 'habitus' can be said to be, like culture, a fluid entity and something individuals have the power to change if they reframe situations and grasp new opportunities to feel 'of use'.

I have already made reference to the significance of entering new spaces, or having one's space invaded as a result of changes of compromised health, as being loss related, but there is evidence that loss can ultimately be transformative (Schneider 2006; Berzoff 2011). And so, becoming dependent on others may not necessarily compromise well-being if it offers new opportunities for reciprocity–friendship, empathic support, the providing of service user feedback and so on. And so, even 'allowing' oneself to be visited can constitute an act of reciprocity if it is perceived as a gift that gives someone pleasure (Lustbader 1991). As a consequence, a sense of mastery can be facilitated.

These have been but a few examples of the many ways in which reciprocity in dependent old age has temporal aspects, some which relate to ontology and some to policy. Though the topics are limited in number and depth of discussion by limitations of space, they provide further evidence that reciprocity is not experienced in a vacuum.

Meaning Making at the Level of the Personal/Spiritual

In this section, I highlight literature that reflects an understanding of the meanings people attach to their experiences of reciprocity (or its neglect) because meaning making is integral to the phenomenological approach to understanding social life, and to my endeavour to give voice to the perspectives of those who use formal care services. The literature can be seen to address two aspects:

- the personal or biographical and
- the cultural or discursive.

My focus in this section is on the former.

The Psychosocial Impact of Meaning Making at the Level of the Personal/Spiritual

The following two points, again just a few examples from many, highlight the significance of meaning making for personal spiritual well-being:

(1) As I have already suggested, for many older people being able to give back contributes to the self-esteem they derive from being what they consider to be 'a good person', in the sense that identity can be said to be not just about what someone 'is' but also what they 'do' (Widdecombe 1998). However, that conceptualisation of 'a good person' is not something that is constructed in isolation from other people's conceptualisation of what constitutes being 'a good person'. That is, for reciprocity to contribute to the construction of a positive self-image, it has to be socially validated and, as we have seen, this is not necessarily the case.

Warr (2005), in referring to poverty, uses the term 'social stigma' in relation to groups that are perceived as being unable to reciprocate for what they receive in terms of support–she highlights single parents, for example. Where older people are in receipt of formal care support, the concept of stigma may well be significant for them if they internalise prevailing messages that they constitute a burden on their families and society in general, and have nothing to offer in compensation. This is proposed by Datta (2008) as being true of Indian elders, where institutional care is often seen as a last resort. But, in discussing the labelling of older women as 'frail', Grenier and Hanley (2007) describe how some older women reject this label through 'acts of resistance' so that their positive self-identities can be maintained:

> Within our studies, older women's resistance to 'frailty' revealed how they experienced the contradictions between powerful social relations, an imposed identity, lived experience and sense of self. On a social level, older women are sometimes caught within the tension between being seen as small, powerless or 'declining into death' (Gadow 1966), and a sense of self that may conflict with these perceptions... 'Frail' older women are not expected to cause problems, make demands or engage in direct action—they are not expected to have access to power and/or resistance. Yet older women resist the concept of 'frailty' on a personal and political level. (p. 217)

Some may be able to engage in communal protest but for others, argue Grenier and Hanley, the acts of resistance are subtle and often go unrecognised as such, partly because they are less overt than the more organised acts of resistance that they are prevented from engaging in by their poor health and the need to 'abide by rules of service eligibility' (p. 223).

There would seem to be parallel here with the recognition, or rather non-recognition, of acts of reciprocity. I would argue that, if there is to be any chance of countering the conceptualisation of dependent older people as a burden, attention has to be paid by those providing formal care to how reciprocity is conceptualised. As with acts of resistance, reciprocity can go unnoticed and unrecognised as such. For example, Beel-Bates et al. (2007) cite older people defining 'pleasantness to staff' and 'gratitude' as examples of reciprocity, which I doubt would be on many people's list if asked to think of examples of reciprocity. Where instances of 'giving back' are not recognised as such, the negative effects on self-esteem are likely to be significant unless a 'virtuous' identity can be constructed by reframing dependency as interdependency (Breheny and Stephens 2009).

(2) The concepts of duty and obligation are also significant in terms of the meaning people attach to reciprocity. For example, inherent in the Hindu faith that the Indian participants in this study followed is an *obligation* to 'give back' to community–a duty (dharma) to reciprocate which has implications that relate to being perceived not only as a good person, but also as having lived a good life for which one may be rewarded in future reincarnations:

> The normative discourses on *dharm* drawing on the conceptualisation of society as a sacred and organic whole, suggest that society is contingent upon everyone performing their sacred duties. (Jacobs 2010, p. 75)

But while the notions of duty and obligation may not be so overtly expressed within other world views, the extent to which they operate at a philosophical level to inform moral codes is something that Gandhi and Bowers (2008) would like to see on a social care agenda that has for too long focused predominantly on demographics at the expense of *understanding* ageing.

Social Capital, Relationships and Meaning Making at the Level of the Personal/Spiritual

The following issues are important in this respect:

(1) Where people lose the ways in which they have experienced 'connectedness' with the world around them, they lose an important aspect of their spirituality. As Attig (2011) comments:

> Our personal relationships and attachments are not accidents but rather essential components of our personal integrity and identity and are decisive in making us who we are. Moreover, our self-concepts, self-images, self-confidence and self-esteem have fundamentally social roots. As individuals we are deeply socially embedded and interdependent in our relationships with others, and our capacities to flourish and to find purpose and meaning in life are social or socially dependent. (p.139)

Affirmation contributes, therefore, to spiritual well-being but we need connectedness in order to experience that affirmation. Research by McMunn et al. (2009) supports the premise that participation in socially productive activities is linked positively to well-being and where relationships prior to becoming dependent on formal care are characterised by 'usefulness' to others, the process of intersubjectivity operates to help us locate ourselves in the social world as good people to know. That is, we have other people's perceptions of us as 'a person with something to offer others' to inform our own view of ourselves. But where one's social network is constituted largely or exclusively by carers (who, according to Bowers et al. 2011, may have a low awareness of the significance of reciprocity in older people's lives) and other dependent people (whose sense of connectedness may also have been challenged) that positive feedback may be hard to come by.

(2) Within childcare literature, there is a significant body of evidence that links resilience—the ability to bounce back from adversity—with positive attachments and effective support networks (Gilligan, 2001; Liebenberg and Ungar, 2008). This association is being explored to a lesser extent, it seems, in gerontological literature. Fuller-Iglesias et al. (2008) claim a positive association between support networks and resilience, and Hildon et al. (2008) identify more particularly the potential for the maintenance of existing valued roles in social networks to contribute to a sense of mastery. Lamb et al. (2009), however, suggest that, while strengths-based approaches (which incorporate resilience) have been shown to be effective in the management of life changes, the literature base of gerontological social work indicates that the care of older people is lagging behind other fields in the integration of this research into practice.

It would seem, then, that if there is an indication that supportive networks (high social capital) may be a factor in helping dependent older people to understand and come to terms with their dependency by focusing on their strengths—particularly in terms of what they still have to give–then a neglect of the association between social capital and resilience may be significant for older people's sense of self-worth and well-being. And, furthermore, for a more sociological understanding of the significance of reciprocity in their lives.

Well-Being and Meaning Making at the Level of the Personal/Spiritual

The following three points highlight more specifically the importance of reciprocity for spiritual well-being:

(1) In terms of well-being, spirituality is an often overlooked, but core concept. As Holloway (2005) comments:

> Spirituality is a notoriously difficult word to define. The word covers the inner life of human beings, all that is left when you have fed and sheltered them, and that's just about everything that's important to them. (p. ix-x)

Her focus on what is 'important to them' chimes well with the sense in which spirituality is concerned with those 'big questions' that we grapple with on our journey through life (Moss 2005), especially when our sense of self is under threat:

- who are we?
- what is our purpose in life?
- do we matter to anyone?

Realising that they have become significantly dependent on others may well be one of those situations when older people feel that their sense of self has been threatened. As their self-esteem is also likely to be under threat, they may internalise the dependency 'label' attached to them, come to accept it as inevitable and live up, or rather down, to the stereotypes associated with that label. Where this happens, it can contribute to the sense in which older people are conceptualised as 'other'–that is, different from 'normal' adults (de Beauvoir 1977, Midwinter 1990). In a magazine article where she refers to her partner's exceptional tallness, a journalist makes a telling comment which serves to highlight how this has become incorporated into common-sense thinking: 'He gets a lot of hugs. From everyone. Men, women, children, the elderly' (Cohen 2009, p. 49). And where 'normal' is not seen to apply, the significance of social engagement and reciprocity may be overlooked as a result.

(2) There is evidence to suggest a positive relationship between social engagement and the limiting of cognitive decline (Keller-Cohen et al. 2006) and with quality of life (Victor et al. 2009). And, in relation to the issue of work ethic as an aspect of engagement, it seems that many older Indians work productively until they are unable to, maybe out of necessity in terms of poverty but perhaps also as part of an ongoing perceived obligation to community, and interdependence, as inherent in Hinduism (Hodge 2004). In that sense, then,

it may be the reciprocity element, not the activity for its own sake, that holds significance for spiritual well-being because of the perceived obligation to have lived a 'good life'.

(3) Within the literature base relating to gift giving, Komter (2007) is among those who highlight its moral dimension, and the sense in which the *meaning* of a gift given is as important as the gift itself, if not more so. Lustbader (1991) claims that dependent older people feel diminished by not being able to reciprocate:

> Frail people are generally denied chances to give something back to their helpers or to their communities. Their offers are refused with statements like, "You do not have to do that. We'll take care of everything." Helpers mean well without realising how urgently people in their care crave a tangible counterbalance to their dependency.(p. 29)

Not recognising, or facilitating, reciprocity can therefore be seen to have the potential to be taken as an affront to dignity in the same way as refusing a more tangible gift has the potential to make someone feel embarrassed or unworthy.

In conclusion, if dependent older people in receipt of formal care are to be able to maintain a sense of themselves as valued people, then their meaning making requires affirmation as such. And where opportunities for reciprocity, and the positive intersubjective feedback it can generate, are not present, then the potential for spiritual well-being to be compromised is high. In the final section of this chapter, I focus more directly on literature that explores *why* those opportunities are not always present.

Meaning Making at the Level of Discourse and Institutionalised Patterns of Power

It is clear, then, that meaning making around reciprocity has a psychological dimension, but its sociological dimension cannot be ignored because, as already discussed, it is clear that human life is social life. Other factors than our own wishes and motivation regarding reciprocity enter the equation in terms of institutionalised patterns of power. By that I mean the social patterns, group relations, shared cultures and so on that have an influence on such matters as what is considered significant as a basis for social division, what life chances are likely to come our way, how we are perceived by others, and so on. I have already touched on the fact that, when we are trying to make sense of the world, and particularly to cope with the many changes (and potential losses) associated with becoming significantly dependent on others, such factors as our age, class and gender (the S level of PCS analysis) are not irrelevant (Arber et al. 2003a, b; Cann and Dean 2009). These matters of social structure are also reflected in, and sustained by, powerful discourses that operate (at the C level of PCS analysis) to persuade us that particular ways of looking at the world make a certain amount of sense. Below, I highlight some of the current discourses that may throw some light on why:

(a) studies of reciprocity beyond a focus on kinship relations, and which focus on those in receipt of formal care are, are limited; and
(b) service user discourses remain subordinate to more dominant ones.

The Psychosocial Impact of Meaning Making at the Level of Discourse and Institutionalised Patterns of Power

The extant literature in this area is broad ranging and vast, and so a comprehensive account is way beyond the scope of this literature review. My intention here is to do no more that reflect some of the sociological literature base that:

(a) challenges the assumption that old age is characterised by homogeneity and
(b) highlights the power of dominant discourses to reinforce the association between old age and decline.

Important considerations include:

(1) That longitudinal studies (Wadsworth 2002; Wenger and Burholt 2004) remind us of the significance of historical or retrospective context for an understanding of the meaning people attach to their experiences of reciprocity or its neglect. This can be seen, for example, in relation to gender. The gendered employment histories of the current cohort of women in advanced old age are likely to have reflected prevailing social expectations about when, where and whether they should have worked in the public sphere, and therefore, the extent which they are valued in the public sphere (Rowbotham 1973). The older men in this cohort are also likely to have constructed their self-image, and acquired a sense of self-esteem, from the role they played earlier in their lives–more likely than not, one that incorporated the notion of men as breadwinners for their families (Price and Ginn 2003). It may be then, that the loss of a former productive or protective role is brought into sharp focus by their reliance on others for help with their daily living.

Studies also provide evidence that, as well as work experience being gendered, retirement also holds different meanings for men and women (Price and Ginn 2003; Kunemund and Holland 2007), and it is likely that these experiences will inform dependent older people's meaning making in connection with reciprocity if they become significantly reliant on others to the extent that existing relationships of give and take become compromised.

(2) Class location may also hold some significance. Acts of reciprocity or 'usefulness' are not dependent on financial means–indeed, as Rakowski et al. (2003) suggest, reciprocity is often conceptualised as being 'in the little things of

daily life', (p. 51), which may not cost much, if anything, in financial terms. However, the financial means to get out and about and engage with others might well have significance for the active engagement in public life from which opportunities for reciprocity might emerge (Victor et al. 2009).

(3) If we return to the concept of 'world-view' we can see that old age and ethnicity also articulate in the sense that there is no single, uncontested social rule about reciprocity–for some cultures the expectation of giving back to community, or to future generations, could be said to be socially validated to the point of being a duty (Rachana Bhangaokar and Shagufa Kapadia 2009; Jacobs 2010), while in others it may be considered peripheral.

(4) I referred in Chap. 2 the power of dominant discourses to portray a particular way of making sense of social life as the *only* way. Several powerful discourses can be seen to reinforce and validate the assumption that old age is characterised only by need, which may help to explain why that assumption is so embedded and resistant to challenge:

bio-medical discourses–those which concern themselves with the body and mastery over it through scientific and medical expertise. The dominant position that medicine continues to hold in terms of epistemological hierarchy helps to make the conceptualisation of old age as illness, or pathology, appear legitimate (Foucault 1973; Powell 2006; Bhaktisvarupta Damodara 2008), thereby helping to embed in the public consciousness the assumption that 'usefulness' in old age is pathological too.

disability discourses–the association of disability with tragedy, charity and the care model has been strongly challenged by the social model of disability which locates the cause of disability in disabling environments and discriminatory social attitudes (Swain et al. 2004; Oliver 2009; Barnes and Mercer 2010) rather than in the perceived deficits of individuals. However, while such discourses have been effective in highlighting dependency as a social construction, the additional dynamic of ageist discourses has worked to limit the extent to which disabled *older* people are considered in disability discourses (Boerner and Rheinhardt 2003). That is, when disability in old age is theorised, it is *age* discourses, rather than *disability* discourses that tend to be drawn on. As a result, the disability movement's focus on social justice may not necessarily be seen to have relevance and, as with bio-medical discourses, this may act as an impediment to reciprocity when older people become dependent on formal care.

citizenship discourses–understanding and addressing the diminishing of citizenship in old age has been a major feature of the critical gerontology agenda, as evident in the following comment by Bytheway (2011):

> … the extensive work of critical gerontologists over the last 30 years has contributed to our understanding of how the idea of old age fits into contemporary ideologies and inter-generational tensions. Townsend's theory of structured dependency (1981), for example, helps us understand better the backdrop rather than the course of individual lives, and to appreciate that people are 'made' old and dependent primarily by well-meaning caring policies and practices, rather than by physiological change. So, more generally, age can be recognised, not only in the life and identity of the individual person, but also in the relationship between people, in social groups and in some of the divisions in society. (p. 215)

And where older people are constructed as not only different, but 'less than', it lends legitimacy to their being treated less favourably in terms of citizenship rights to, for example, dignity and respect. The relative exclusion of older people from citizenship debates may be at least partly explained by;

(a) the enduring influence of the bio-medical paradigm to sustain the 'welfarisation' (Fennell et al. 1988) of older people and
(b) the extent to which their marginalisation from the economic sphere (only recently and modestly being addressed in the UK) has been sustained by ageist assumptions about the superiority of youth.

As such, their situation is well described by Midwinter's use of the term 'post-adulthood' (Midwinter 1990). As with the marginalisation of older people within disability discourses, marginalisation within citizenship discourses serves to dilute the potential for citizenship discourses to challenge the dominance of the bio-medical discourses which question why the issue of 'giving back' in old age should be on the research agenda at all.

Social Capital, Relationships and Meaning Making at the Level of Discourse and Institutionalised Patterns of Power

The dominance of a 'care' model helps to legitimise the confining of older people who need extra support to 'places of care'–be that in their own homes or in institutions. While attention has been paid in the research field to the combating of isolation in old age (Walker 2005; Victor et al. 2009; Windle et al. 2011), and the encouraging of participation in civic engagement and decision making in general (Bowers et al. 2009; Anderson and Dabelko-Schoeny 2010), addressing the marginalisation of dependent older people–be it in ideological or practical terms–does not appear to be high on political agenda at present. From a policy point of view in both the UK and India, the concern appears to be focused much more on where, and at a cost to whom, people will be 'looked after', than on how the means to continue living preferred lifestyles, in preferred environments (with their existing network of reciprocal relationships) can be facilitated, though attention to assistive technologies may be beginning to address this.

For as long as dependency in old age is necessarily associated with illness and decline, a 'care' model is likely be considered to be appropriate–and if older people who need help continue to be defined *only* in relation to being 'in need', then care provision as a one-way process of being 'done to' is likely to continue to be seen as appropriate (Boerner and Rheinhardt 2003; Bowers et al. 2011). As a consequence, opportunities to build up the social capital associated with positive affirmation is likely to remain low, and spiritual well-being compromised as a result.

Well-Being and Meaning Making at the Level of Discourse and Institutionalised Patterns of Power

And finally, a brief exploration of some of the broader factors that have significance for the potential of reciprocity to contribute to well-being:

(1) A consideration of the association between well-being and institutionalised patterns of power begs the question of which aspect of well-being we are talking about. My particular concern is for the spiritual well-being of people in receipt of formal care but, given that societies are characterised by competing interests, and that a number of enduring dominant discourses combine to sustain the assumption that dependent older people do not have a spiritual dimension–and therefore no need for affirmation–then it comes as no surprise that spiritual needs tend not to be well addressed (Canda and Furman 1999; Moss 2005).

I would not deny that older people in care situations are without their champions altogether, but those initiatives I have encountered ('A Dignified Revolution', www.dignifiedrevolution.org.uk; the Dignity Foundation, *Dignity Dialogue*, 2009) tend to focus more heavily on issues relating to the body and personal care, than on the reciprocity that could contribute to well-being.

Furthermore, one of the key points to emerge from an evaluation by Blood (2010) of accounts of research (undertaken by the Joseph Rowntree Foundation) into the quality of life of older people with high support needs, is that there is limited evidence of a service user perspective. From a well-being point of view, I would suggest that having one's meaning making considered epistemologically inferior to that of care providers and researchers could constitute an affront to one's sense of self-worth and therefore to be potentially detrimental to well-being. It seems, then, that without a reconceptualisation of old age, and older people's value, their significant existing and future potential as contributors to society, is unlikely to be championed at the level where policy change could make a significant difference to older people's lives.

In challenging the correlation between happiness and wealth, Jordan (2008) highlights relationship building and social participation as crucial to well-being and quality of life, but comments that it is missing from current social policy in the UK. In such a climate it is hard to imagine a reconceptualisation of the '4th age' occurring unless the change comes about from the bottom up–from older people themselves becoming the agents of change. However, for those already compromised in terms of social networks, this is not likely to be easy.

(2) As with the social construction of disability referred to earlier, it has been argued that vulnerability too can be socially constructed (Martin 2007). That is, it is not inevitable. For example, not wrapping people up in cotton wool can allow them the chance to develop new, or enhance existing, coping skills. Furthermore, facilitating the risk taking that forms part of everyday life, rather than denying it, can enhance, rather than compromise, well-being (Lloyd 2006). This may therefore

have implications for the extent to which opportunities for reciprocity are seen to be appropriate for those older people in receipt of formal care.

(3) As I have already alluded to, Indian elders traditionally tend not to be seen as vulnerable just because they are old (Nair 2008). This may be related to their living, for the most part, within communities and families where what older people can contribute is visible to the wider population. With a rise in the number of care homes as a response to abandonment or lifestyle choice (Srikrishna 2006), older people themselves and their 'usefulness to others' may become increasingly invisibilised.

If the 'giving back' from which people are assumed to derive self-esteem (such as through volunteering, engaging in activism, peer counselling and so on) have the potential to contribute to spiritual well-being by validating people's contributions, then meaning making at this level is concerned with why such opportunities are not facilitated by those providing or managing the formal care of dependent older people, and why dependent older people have to try and make sense of their lives in the face of messages writ high, large and often that they are just bodies, rather than *people* with feelings, and aspirations.

Conclusion

From this broad, sweep of extant literature relating to reciprocity and eldercare I would highlight the following findings as representative of the current research and scholarship on which my study builds. It is already indicated that:

- staying active and engaged has positive benefits for both physical and mental well-being;
- having supportive networks and high reserves of social capital helps older people to cope with adversity;
- older people need affirmation as valued people if their spiritual health is not to be diminished;
- this affirmation is not readily available when older people become isolated from the communities or situations in which they had previously held valued roles;
- life is experienced in changing temporal and spatial contexts which have significance for how dependency and reciprocity are both experienced and conceptualised and
- dominant discourses help to construct and maintain these conceptualisations of old age as needy or otherwise, particularly through processes of stereotyping (Pickering 2001).

Less appears to be known about how reciprocity, or its neglect, is experienced from service user and meaning making perspectives. For example, when reciprocity has been a feature of the earlier years of dependent older people's lives, such that it had contributed to their affirmation as valued people, how does it feel when it is neglected or denied? If doing so causes spiritual or existential discomfort by challenging their sense of self, then how do they make sense of that? The phenomenological study that I now go on to present in Part Two builds on the existing research I have identified here, to explore these questions, and the part that the dialectical processes inherent in PCS analysis play in accounting for why, and how, reciprocity in eldercare is impeded or promoted.

References

Anderson, K. A., & Dabelko-Schoeny, H. J. (2010). Civic engagement for nursing home residents: A call for social work action. *Journal of Gerontological Social Work, 53*, 270–282.
Andrews, G. J., Gavin, N., Begley, S., & Brodie, D. (2003). Assisting friendships, combating loneliness: Users' views on a 'befriending' scheme. *Ageing and Society, 23*(3), 349–362.
Antaki, C., & Widdecombe, S. (Eds.). (1998). *Identities in talk*. London: Sage.
Arber, S., Davidson, K., & Ginn, J. (2003). *Changing approaches to gender and later life*. In Arber et al. (Eds.), Maidenhead: Open University Press.
Arber, S., Davidson, K., & Ginn, J. (Eds.). (2003b). *Gender and ageing: Changing roles and relationships*. Maidenhead: Open University Press.
Attig, T. (2011). *How we grieve: Relearning the world*. New York: Oxford University Press. revised edn.
Barnes, C., & Mercer, G. (2010). *Exploring disability* (2nd ed.). Cambridge: Polity Press.
de Beauvoir, S. (1977). *Old age*. Harmondsworth: Penguin.
Beel-Bates, C. A., Ingersoll-Dayton, B., & Nelson, E. (2007). Deference as a form of reciprocity among residents in assisted living. *Research on Aging, 29*, 626–643.
Bernard, M., & Scharf, T. (Eds.), (2007). *Critical perspectives on ageing societies*. Bristol: The Policy Press.
Berzoff, J. (2011). The transformative nature of grief and bereavement. *Clinical Social Work Journal, 39*(3), 262–269.
Bhaktisvarupta Damodar, H. H. (2008). *Aging and dying: The vedantic perspective*. In Chatterjee et al. (Eds.), Oriental disadvantage versus occidental exuberance. *International Sociology, 23*(1), 5–33.
Bianchi, E. (2005). Living with elder wisdom. *Journal of Gerontological Social Work, 45*(3), 319–329.
Blood, I. (2010). *Older people with high support needs: How can we empower them to enjoy a better life?*. York: Joseph Rowntree Foundation.
Boerner, K., & Reinhardt, J. P. (2003). Giving while in need: Support provided by disabled older adults. *Journal of Gerontology, 58B*(5), S297–S304.
Bond, J., Peace, S., Dittman-Kohli, F. & Westerhof, G. (Eds.), (2007). *Ageing in society: European perspectives on gerontology* (3rd ed.). London: Sage.
Bornat, J., & Bytheway, B. (2010). Perceptions and presentations of living with risk in everyday life. *British Journal of Social Work, 40*, 1118–1134.
Bornat, J., & Tetley, J. (2011). *Oral history and ageing*. London: CPA.
Bourdieu, P. (1986). *The forms of capital*. In Richardson (Ed.), *The handbook of theory and research for the sociology of education*. New York: Greenwood Press.
Bourdieu, P. (1984). *Distinction: Social critique of the judgement of taste*. Cambridge: Harvard University Press.
Bowers, B. J., Fibich, B., & Jacobson, N. (2001). Care-as-service, care-as-relating, care-as-comfort: Understanding nursing home residents' definitions of quality. *The Gerontologist, 41*(4), 539–545.
Bowers, H., Clark, A., Crosby, G., Easterbrook, L., Macadam, A., Macdonald, R., Macfarlane, A., Maclean M., Patel., M., Runnicles, D., Oshinaike, T., & Smith, C. (2009). *Older people's vision for long-term care*. York: Joseph Rowntree Foundation.

References

Bowers, H., Mordey, M., Runnicles, D., Barker, S., Thomas, N., Wilkins, A., et al. (2011). *Not a one-way street: research into older people's experiences of support based on mutuality and reciprocity*. Joseph Rowntree Foundation: York.

Bradley, N., & Popper, W. (2003). Assistive technology, computers and Internet may decrease sense of isolation for homebound elderly and disabled persons. *Technology and Disability, 15*, 19–25.

Breheny, M., & Stevens, C. (2009). "I sort of pay back in my own little way": Managing independence and social connectedness through reciprocity. *Ageing and Society, 29*, 1295–1313.

Bury, M. (1982). Chronic illness as biographical disruption. *Sociology of Health & Illness, 4*, 165–182.

Bytheway, B. (2011). *Unmasking age: The significance of age for social research*. Bristol: The Policy Press.

Canda, E., & Furman, L. (1999). *Spiritual diversity in social work practice: The heart of helping*. New York: The Free Press.

Cann, P., & Dean, M. (Eds.), (2009). *Unequal ageing: The untold story of exclusion in old age*. Bristol: The Policy Press.

van Castiglione, D., Deth, J. W., & Wolleb, G. (Eds.), (2008). *The handbook of social capital*. Oxford: Oxford University Press.

Cattan, M. (Ed.), (2009). *Mental health and well-being in later life*. Maindenhead: Open University Press.

Chatterjee, D. P. (2008). Oriental disadvantage versus occidental exuberance. *International Sociology, 23*(1), 5–33.

Chokkanathan, S. (2009). Resources, stressors and psychological distress among older adults in Chennai, India. *Social Science and Medicine, 68*, 243–250.

Clow, A., & Aitchison, L. (2009). *Keeping active*. In Cattan (Ed.), *Mental health and well-being in later life*. Maindenhead: Open University Press.

Cohen, G. D. (2005). *The mature mind: The positive power of the aging brain*. New York: Basic Books.

Cohen, A. (2009) Boys ignored me. My closest resemblance was to Big Bird. *The Week*, (20th June 2009).

Coleman, J. (1988). *Foundations of social theory*. Cambridge: Harvard University Press.

Commission for Social Care Inspection. (2006). *Time to care?*. Newcastle: An overview of home care services for older people in England. CSCI.

Coote, A. (2009). *The uneven dividend: Health and well-being in later life*. In Cann and Dean (Eds.), *Unequal ageing: The untold story of exclusion in old age*. Bristol: The Policy Press.

Crossley, N. (1996). *Intersubjectivity: The fabric of social becoming*. London: Sage.

Crossley, N. (2005). *Key concepts in critical theory*. London: Sage.

Datta, A. (2008). *Socio-ethical issues in the existing paradigm of care for the older persons: Emerging challenges and possible responses*. In Chatterjee et al. (Ed.)

Davidson, K., Warren, L., & Maynard, M. (2005). *Social involvement: Aspects of gender and ethnicity*. In Walker (Ed.), *Dignity dialogue* (January 2009) Mumbai: Dignity Foundation.

Dignity Dialogue (January, 2009) Mumbai, Dignity Foundation

Edwards, R., & Glover, J. (Eds.), (2001). *Risk and citizenship: Key issues in welfare*. Abingdon: Routledge.

Fennell, G., Phillipson, C., & Evers, H. (1988). *The sociology of old age*. Milton Keynes: Open University Press.

Fennema, M., & Tillie, J. (2008). *Social capital in multicultural societies*. In Castiglione et al. (Eds.), Oxford: Oxford University press

Forte, D. (2009). *Relationships*. In Cattan (Ed.), *Mental health and well-being in later life*. Maindenhead: Open University Press.

Foucault, M. (1973). *The birth of the clinic*. London: Routledge.

Foucault, M. (1977). *Discipline and punish: The birth of the prison*. London: Allen Lane.

Fromm, E. (1955). *The sane society*. New York: Rinehart.

Fuller-Iglesias, H., Sellars, B., & Antonucci, T. C. (2008). Resilience in old age: Social relations as a protective factor. *Human Development, 5*(3), 181–193.

Gadow, S. (1966). Aging as death rehearsal: The oppressiveness of reason. *The Journal of Clinical Ethics, 7*(1), 35–40.
Galit, N. (2010). Seniors' online communities: A quantitative content analysis. *Gerontologist, 50*(3), 382–392.
Gandhi, K., & Bowers, H. (2008). *Duty and obligation: The invisible glue in services and support*. York: The Joseph Rowntree Foundation.
Garey, A. I., Hansen, K. V., Hertz, R., & Macdonald, C. (2002). Care and kinship: An introduction. *Journal of Family Issues, 23*, 703–715.
Garrett, P. M. (2010). Making social work more Habermasian? A rejoinder in the debate on Habermas. *The British Journal of Social Work, 40*(6), 1754–1758.
Gergen, K. J. (2009). *An invitation to social construction* (2nd ed.). London: Sage.
Gibson, F. (2011). *Reminiscence and life story work* (4th ed.). London: Jessica Kingsley Publishers.
Giddens, A. (2009). *Sociology* (6th ed.). Cambridge: Polity Press.
Gilligan, R. (2001). *Promoting resilience: A resource guide on working with children in the care system*. London: BAAF.
Glanville, J. L., & Bienenstock, E. J. (2009). A typology for understanding the connections among different forms of social capital. *American Behavioural Scientist, 52*(11), 1507–1530.
Godfrey, M., Townsend, J., & Denby, T. (2004). *Building a good life for older people in local communities*. Joseph Rowntree Foundation: York.
Goffman, E. (1961). *Asylums*. Harmondsworth: Penguin.
Gouldner, A. W. (1960). The norm of reciprocity: A preliminary statement. *American Sociological Review, 25*(2), 161–178.
Gray, A. (2009). The social capital of older people. *Ageing and Society, 29*, 5–31.
Grenier, A., & Hanley, J. (2007). Older women and "frailty": Aged, gendered and embodied resistance. *Current Sociology, 55*, 211–228.
Groombridge, B. (1982). Learning, education and later life. *Adult Education, 54*, 314–325.
Gulette, M. M. (2004) *'The Sartre-de Beauvoir "conversations" of 1974': From life storytelling to age autobiography*. In Johnson (Ed.). *Writing old age*. London: Centre for Policy on Ageing.
Hafford-Letchfield, T. (2010). 'The age of opportunity? Revisiting assumptions about the lifelong learning opportunities of older people using social care services. *British Journal of Social Work, 40*, 496–512.
Hall, S., Longhurst, S., & Higginson, I. J. (2009). Challenges to conducting research with older people living in nursing homes. *BMC Geriatrics, 9*(38), 1–6.
Harrington Meyer, M. (2000). *Care work: Gender class and the welfare state*. London: Routledge.
Harvey, J. H. (Ed.), (1998). *Perspectives on Loss: A Sourcebook*. Philadelphia: Brunner Routledge.
Hatton-Yeo, A. (Ed.), (2006). *Intergenerational programmes: An introduction and examples of practice*. Stoke-on-Trent: Beth Johnson Foundation.
Heidegger, M. (1962 trans.) *Being and time: A translation of Sein und Zeit*. New York: University of New York Press.
Help the Aged. (2007). *My Home Life, Issue 1*. London: Help the Aged.
Hildon, Z., Smith, G., Netuveli, G., & Blane, D. (2008). Understanding adversity and resilience at older ages. *Sociology of Health & Illness, 30*(5), 726–740.
Hiller, H. H., & Franz, T. M. (2004). New ties, old ties and lost ties: the use of internet in diaspora. *New Media Society, 6*, 731–752.
Hodge, D. R. (2004). Working with Hindu clients in a spiritually sensitive manner. *Social Work, 49*(1), 27–38.
Holland, C., Kellaher, L., Peace, S., Scharf, T., Breeze, E., Gow, J., & Gilhooly, M. (2005) *Getting out and about*. In Walker (Ed.), *Understanding quality of life in old age*. Maidenhead: Open University Press.
Holloway, M. (2005). *Looking in the distance*. New York: Canongate.

Jacobs, S. (2010). *Hinduism today*. London: Continuum.
Jamieson, A., & Victor, C. R. (Eds.), (2002). *Researching ageing and later life*. Buckingham: Open University Press.
Janoff-Bulman, R. & Berg, M. (1998). Disillusionment and the creation of values. In Harvey (Ed.), *Perspectives on Loss: A Sourcebook*. Philadelphia: Brunner Routledge.
Johnson, J. (Ed.) (2004). *Writing old age*. London: Centre for Policy on Ageing.
Johnson, A. (1998, Nov/Dec) All play and no work? Take a fresh look at activities. *Journal of Dementia Care, 6*(6), 25–27.
Jones, B., & Miller, B. (2001). *Social capital and waves of innovation in risk society*. In Edwards and Glover (Eds.), *Risk and citizenship: Key issues in welfare*. Abingdon: Routledge.
Jordan, B. (2008). *Welfare and well-being: Social value in public policy*. Bristol: The Policy Press.
Keller-Cohen, D., Fiori, K., Toler, A., & Bybee, D. (2006). Social relations, language and cognition in the "oldest old". *Ageing and Society, 26*, 585–605.
Knight, T., & Mellor, D. (2007). Social inclusion of older adults in care: Is it just a question of providing activities? *International Journal of Qualitative Studies in Health and Well-being, 2*(2), 74–85.
Komter, A. (2007). Gifts and social relations: The mechanisms of reciprocity. *International Sociology, 22*, 93–106.
Kunemund, H., & Holland, F. (2007). *Work and retirement*. In Bond et al. (Eds.), *Ageing in society: European perspectives on gerontology* (3rd ed.). London: Sage.
Lamb, F. F., Brady, E. M., & Lohman, C. (2009). Lifelong resiliency learning: A strengths-based synergy for gerontological social work. *Journal of Gerontological Social Work, 52*, 713–728.
Liebenberg, L., & Ungar, M. (Eds.), (2008). *Resilience in action*. London: University of Toronto Press.
Lin, N. (2008). *A network theory of social capital*. In Castiglione et al. (Eds.), pp 50–69. New York: Oxford University Press.
Lloyd, M. (2002). *A framework for working with loss*. In Thompson (Ed.). Basingstoke: Palgrave Macmillan.
Lloyd. M. (2006). Let people loose. Community Care, 26th November.
Lustbader, W. (1991). *Counting on kindness: The dilemmas of dependency*. London: The Free Press.
Mackinlay, E. (2001). *The spiritual dimension of ageing*. London: Jessica Kingsley Publishers.
Manthorpe, J., Cornes, M., Rapaport, J., Moriarty, J., Bright, L., Clough, R., et al. (2006). Commissioning community well-being: focus on older people and transport. *Journal of Integrated Care, 14*(4), 28–37.
Martin, J. (2007). *Safeguarding adults*. Lyme Regis: Russell House Publishing.
Maya-Jariego, I., & Armitage, N. (2007). Multiple senses of community in migration and commuting: The interplay between time, space and relations. *International Sociology, 22*(6), 743–766.
McKee, K. J. (1998). The body drop: A framework for understanding recovery from falls in older people. *Generations Review, 8*, 101–108.
McMunn, A., Nazroo, J., Wahrendorf, M., Breeze, E., & Zaninotto, P. (2009). Participation in socially-productive activities, reciprocity and well-being in later life: Baseline results in England. *Ageing and Society, 29*, 765–782.
Means, R. (2007). *The re-medicalisation of later life*. In Bernard and Scharf (Eds.), *Critical perspectives on ageing societies*. Bristol: The Policy Press.
Midwinter, E. (1990). An ageing world: The equivocal response. *Ageing and Society, 10*(2), 221–228.
Moriarty, J., & Butt, J. (2004). *Social support and ethnicity*. In Walker and Hennessy (Eds.), *Growing older: Quality of life in old age*. Maidenhead: Open University Press.
Moss, B. (2005). *Religion and spirituality*. Lyme Regis: Russell House Publishing.
Nair, K. R. G. (2008). *A plea for a holistic approach to aging*. In Chatterjee et al. (Eds.)
Narotzky, S., & Moreno, P. (2002). Reciprocity's dark side: Negative reciprocity, morality and social reproduction. *Anthropological Theory, 2*, 281–305.

Natarajan, C. N. (2000). *Approach to the aged: A critique*. In Sudhir (Ed.), *Aging in rural India: Perspectives and prospects*. Delhi: Indian Publishers.
Neimeyer, R. A., & Anderson, A. (2002). *Meaning reconstruction theory*. In Thompson (Ed.), Basingstoke: Palgrave Macmillan
Neimeyer, R. A., & Sands, D. C. (2011). *Meaning reconstruction in bereavement: From principles to practice*. In Neimeyer et al. (Eds.), *Grief and bereavement: Bridging research and practice*. London: Routledge.
Neimeyer, R. A., Harris, D. L., Winokuer, H. R., & Thornton, G. F. (Eds.). (2011). *Grief and bereavement: Bridging research and practice*. London: Routledge.
Nowell-Smith, G. (Ed.), (1998). *Antonio Gramsci: Selections from the Prison Notebooks*. London: Lawrence and Wisehart.
Oliver, M. (2009). *Understanding disability: From theory to practice* (2nd ed.). Palgrave Macmillan: Basingstoke.
Palmore, E. (2000). Ageism in gerontological language. *Gerontologist, 40*(6), 645.
Peace, S., Holland, C., & Kellaher, L. (2006). *Environment and identity in later life*. Maidenhead: Open University Press.
Peace, S., Wahl, H-W., Mollenkopf, H., & Oswald, F. (2007). *Environment and Ageing*. In Bond et al (Eds.), *Ageing in society: European perspectives on gerontology* (3rd ed.). London: Sage.
Pickering, M. (2001). *Stereotyping: The politics of representation*. Palgrave Macmillan: Basingstoke.
Poon, L. W., Gueldner, S. H., & Sprouse, B. M. (Eds.), (2003). *Successful ageing and adaptation with chronic disease*. New York: Springer.
Powell, J. L. (2006). *Social theory and ageing*. Oxford: Rowman and Littlefield.
Price, D., & Ginn, J. (2003). Sharing the crust?: *Gender, partnership status and inequalities in pension accumulation*. In Arber et al. (2003).
Putnam, R. D. (2000). *Bowling alone: The collapse and revival of American community*. New York: Simon and Schuster.
Rachana Bhangaokar and Shagufa Kapadia. (2009). 'At the interface of 'Dharma' and 'Karma': Interpreting moral discourse in India. *Psychological Studies, 54*(2), 96–108.
Rakowski, W., Clark, M. A., Miller, S., & Berg, K. M. (2003). *Successful ageing and reciprocity among older adults in assisted living settings*. In Poon et al. (Eds.), New York: Springer.
Raske, M. (2010). Nursing home quality of life: Study of an enabling garden. *Journal of Gerontological Social Work, 53*, 336–351.
Renzenbrink, I. (2004). Home is where the heart is. *Illness, Crisis and Loss, 12*(1), 63–74.
Richardson, J. (1986). *The handbook of theory and research for the sociology of education*. New York: Greenwood Press.
Rowbotham, S. (1973). *Woman's consciousness man's World*. Harmondsworth: Penguin.
Ryvicker, M. (2009). Preservation of self in the nursing home: contradictory practices within two models of care. *Journal of Aging Studies, 23*, 12–23.
Sartre, J.-P. (1963). *Search for a method*. New York: Vintage.
Schneider, J. M. (2006). *Transforming loss: A discovery process*. Michigan: Integra Press.
Shachter-Shalomi, Z., & Miller, R. S. (1997). *From age-ing to sage-ing: A profound new vision of growing older*. New York: Grand Central Publishing.
Social Care Institute for Excellence. (2009). *Personalisation briefing: Implications for home care providers*. London: SCIE.
Srikrishna, L. (2006). Old age homes "preferred by many elders". *The Hindu*, 25th November.
Stanton, G., & Tench, P. (2003) Intergenerational storyline bringing the generations together in North Tyneside. *Journal of Intergenerational Relationships, 1*(1), 71–80.
Sudhir, M. A. (Ed.). (2000). *Aging in rural India: Perspectives and prospects*. Delhi: Indian Publishers.
Swain, J., French, S., Barnes, C., & Thomas, C. (2004). *Disabling barriers: Enabling environments* (2nd ed.). London: Sage.
Tanner, D. (2010). *Managing the ageing experience: Learning from older people*. Bristol: The Policy Press.

References

Thompson, N. (1998). The ontology of ageing. *The British Journal of Social Work,* 28(5), 695–707.
Thompson, N. (Ed.), (2002a). *Loss and grief.* Palgrave Macmillan: Basingstoke.
Thompson, N., & Thompson, S. (2001). Empowering older people: Beyond the care model. *Journal of Social Work,* 1(1), 61–76.
Thompson S. (2002). *Older People.* In Thompson (Ed.), *Loss and grief.* Palgrave Macmillan: Basingstoke.
Tournier, P. (1972). *Learning to grow old.* London: SCM Press.
Uehara, E. S. (1995). Reciprocity reconsidered: Gouldner's "moral norm of reciprocity" and social support. *Journal of Social and Personal Relationships,* 2, 483–502.
Ungerson, C. (2000). *Cash in care.* In Harrington Meyer (Ed.), *Care work: Gender class and the welfare state.* London: Routledge.
Victor, C., Scambler, S., & Bond, J. (2009). *The social world of older people: Understanding loneliness and social isolation in later life.* Maidenhead: Open University Press.
Wadsworth, M. (2002). *Doing longitudinal research.* In Jamieson & Victor (Eds.), *Researching ageing and later life.* Buckingham: Open University Press.
Walker, A. (Ed.). (2005). *Understanding quality of life in old age.* Maidenhead: Open University Press.
Walker, A., & Hagan Hennessy, C. (Eds.), (2004). *Growing older: Quality of life in old age.* Maidenhead: Open University Press.
Warr, D. J. (2005). Social networks in a discredited community. *Journal of Sociology,* 41, 285–308.
Warren, M. E. (2008). *The nature and logic of bad social capital.* In Castiglione et al. (Eds.)
Wenger, G. C., & Burholt, V. (2004). Changes in levels of social isolation and loneliness among older people in a rural area: A 20 year longitudinal study. *Canadian Journal on Aging,* 23(2), 115–127.
Whittall, D., & Grace, D. (2002). Bridging the technology gap: Enhancing communication between residents in a nursing home with local authority school children using the internet. *Geriaction,* 20(4), 15–17.
Widdecombe, S. (1998). *Identity as an analyst's and a participant's resource.* In Antaki & Widdecombe (Eds.), *Identities in talk.* London: Sage.
Windle, K., Francis, J., & Coomber, C. (2011). *'Preventing loneliness and social isolation: Interventions and outcomes', SCIE Research Briefing 39.* London: SCIE.

Part II
Hearing Their Voices

Chapter 4
Research Design and Methods

Introduction

In the previous chapter, I discussed the broad context in which this research is located and, in doing so, addressed the rationale for having a phenomenological approach underpin design and methods. What follows is a discussion of the methodological issues pertinent to this particular piece of empirical research—the application in the field of the methodological issues discussed previously in terms of the phenomenological paradigm. In the first part of this chapter, I focus on the research design, outlining my rationale for the choice of research methods that I discuss later. In describing and justifying my choices as they relate to sampling, interview schedule and analytical tool, I would highlight three sets of issues as being of particular significance because, to my mind, and in the opinions of researchers mentioned below, they are fundamental to good-quality research.

The first is that of academic rigour. I discuss in more detail below how I claim this study to be academically rigorous but would make the general point here that I agree with Padgett (1998) that, while academic rigour and trustworthiness are necessary if research is to have credibility, adhering to positivist research benchmarks such as generalisabilty and replicability can obscure the aims of qualitative research:

> Other strategies for scientific rigor do not easily apply to qualitative methods, including *random sampling, generalizability*, and *reliable and valid measurement… generalizability* is not a priority in qualitative studies, where the uniqueness of the human experience is celebrated. (p. 91, emphasis in original)

The second set of issues relate to research ethics, particularly as they apply to gerontology. While I explore ethical issues specific to particular aspects of design and method in the sections that follow, I want here to touch on a more general concern about undertaking research with what many people consider to be a particularly vulnerable group of people. Albeit often in the interests of promoting the well-being of a larger population, research has the potential to compromise the well-being of those chosen to represent that population (be that through acts of omission, such as lacking awareness of people's fears or rights, or commission, by causing distress or emotional harm). I have sought to guard against compromising the well-being of those who told me their stories but, while I agree with Holland (2005) that there is considerable potential for frail older people to be harmed or inconvenienced by being 'over-researched', the potential also exists for participation in research activity to have positive, and even transformational, outcomes for older participants if they are able to reframe their self-image as a consequence. As Holland (ibid) and Peace (2002) point out, the group in question is one from which competent research-active partners have the potential to emerge as new ways of involving them are developed.

While I do accept that, in some respects, the people I interviewed could be described as vulnerable—indeed they were chosen as representative of a population who have been deemed so by the eligibility criteria of welfare providers—I did not see this as problematic in terms of their playing an active role in research activity if they were willing to do so and I could facilitate it. I considered that, having had many years of experience of working with older people as a nurse and social worker, I possessed the skill and sensitivity to put them at their ease in an interview situation, and to be alert and responsive to any signs of distress or discomfort that might have arisen. And while I did not articulate it to them in those terms, I considered the role that the participants in my study played as exemplifying the very focus of it—reciprocity. That is, an opportunity to give to others—to have a 'useful' role in the sense of knowing that the partnership of their insights, and my analysis and disseminating of them, had the potential to inform, and maybe improve, the care of their peers and future cohorts of dependent older people.

Concerned as I am to operate as what Roulston (2010) refers to as a 'mindful' researcher by both exploring and making explicit my values, theoretical, ontological and epistemological stances, I kept a reflective journal throughout the processes of planning, executing and recording my research. Conversations with myself, such as those relating to the researcher-researched relationship have featured in the journal as examples of the third key set of issues in this chapter—those relating to reflexivity. In order for readers to have as sound a basis as possible for assessing the academic rigour of my study, my aim has been to be as transparent as I can be about how I position myself as a researcher in the social world in which I operate, and how I recognise and address the possibilities for bias in my work.

The Topic: Why Reciprocity?

My decision to focus on reciprocity was borne out of many years of working with frail older people as both a nurse and a social worker, and noticing how frequently they are conceptualised by those who organise or provide their care as being in *need* of help but not also able to *give* help or contribute to the well-being of others. Working on the premises that even those people who have very limited physical or mental capacity have the potential to reciprocate in some way, and that a sense of 'connectedness' contributes towards our self-esteem and therefore general well-being (Antonucci and Jackson 1989; Graham 2007; Jordan 2008; Attig 2011), I am concerned, as are Bowers et al. (2011), that the concept tends to be either absent from, or only peripheral to, policy and practice agenda relating to old age in general, and the promotion of dignity in particular. If we consider that we all live in the world as social beings, then to deny older people opportunities to interact, and be validated within it, is to deny them their humanity—to conceptualise them as 'less than' other sectors of the population (de Beauvoir 1997; Midwinter 1990). And where this challenges dependent older people's biographical continuity—their sense of 'staying me'—then the potential for the estrangement from self-referred to above (Fromm 1955) would seem to be heightened, and their spiritual well-being compromised as a result.

This study is premised on the argument that reciprocity has a key role to play in promoting the well-being of dependent older people (defined in this study as being in receipt of formal care—that is, care beyond that provided by family) at two levels:

- It is significant at an individual level because it is one of the factors which contributes to identity formation. It has the potential to reinforce a positive identity which is linked to high self-esteem, confidence and self-empowerment but, on the other hand, also to reinforce low self-esteem and self-confidence, which increases the likelihood that the constant drip-feed of ageist messages that older people are subjected to (Chasteen et al. 2002; Whitbourne and Sneed 2002) will become internalised by them.
- In terms of social well-being, it is significant because it has a role to play in promoting social justice and social inclusion by highlighting that older people, whatever their level of ability and competence as citizens, contribute to society's resources as well as draw from them.

To neglect the 'social' aspect of their status as social beings is to deny them a framework for understanding and making sense of their status and role in the world in which they continue to operate, however compromised that might become by ill-health or incapacity. In effect, it has the potential to give the message: 'we're not interested in you or how you feel about yourself'. My study aims to give voice to older people who have become substantially dependent on others for help with their daily living, by exploring whether the skills and knowledge that underpinned their capacity to operate as reciprocating citizens earlier in their lives are still recognised and valued in their current situations of dependency.

Keeping reciprocity on the research agenda has the potential to challenge ageist ideologies and contribute towards reframing old age, and the way we treat older people. Furthermore, developing a theoretical framework, as I do, which calls on concepts related to meaning making can provide a tool to aid critically reflective eldercare practice in whatever discipline it operates.

Research Aims

To reiterate, the aims of this study are fourfold:

1. To explore the extent to which reciprocity features in the lives of older people in receipt of formal care, I seek a service user perspective on whether:

    (a) they have experienced a mismatch between the image they hold of themselves as someone who can reciprocate, and prevailing negative stereotypes of dependent older people;

(b) opportunities to reciprocate—that is, to continue operating as the person they have always considered themselves to be—have been facilitated or denied by care providers; and
(c) they have experienced a sense of alienation or existential discomfort as a result.

2. To analyse the data produced with regard to similarities and differences in terms of experience of, and opportunities for, reciprocity.
3. To theorise the data produced so that those 'insider' stories become more than just a collection of individual accounts. In her feminist critique of standpoint theory, which has at its root the assumption that no general claims can be made from individual experience because individual perspectives or 'standpoints' are unique, Bubeck (2000) proposes a dialogical model to challenge those who focus on difference and antagonism:

> In conclusion, the dialogical model allows us to see how commonality can become a possibility in the face of acknowledged difference. Given dialogical processes, the biases and distortions produced by oppressive conditions can be counteracted and overcome: difference does not have to be eternally divisive, nor do oppressive conditions have to be perpetuated forever. (pp. 200–201)

Given the phenomenological basis of my study, I feel that I can assume the uniqueness of experience as a given, but agree with Bubeck that the search for commonality of experience is a legitimate endeavour.

4. Having listened to, analysed and theorised the accounts of individual experience within an analytical framework that takes account of the cultural and structural contexts of individual experience, to propose a theoretical framework for informing eldercare practice of the significance of reciprocity in the lives of dependent older people.

Research Focus

It will be evident from the discussions in previous chapters that my analysis of current thinking relating to reciprocity and eldercare has been informed by the recognition that life is not experienced by individuals in a vacuum, but within cultural and structural contexts (Rubenstein 2001; Giddens 2009; Thompson 2011). My analysis of the literature has also been informed by phenomenological concepts that relate to social life and people's perceptions and expectations within it. To reiterate, I have called on four concepts that are key to the phenomenological paradigm:

- *social space*—concerned as it is with relationships;
- *temporality*—relating to social time as a significant feature of continuing biography;
- *meaning making at the level of personal and spiritual concerns*; and
- *meaning making at the level of discourse and institutionalised patterns of power*.

As these concepts, and my rationale for choosing them as a framework for analysing both the literature and data, have already been explored, I will not be revisiting that discussion here, except to say that they have formed the basis of the research questions I set myself as follows:

(1) In what ways is the concept of social space significant in accounting for how reciprocity in the care of dependent older people in receipt of formal care is promoted or impeded?
(2) In what ways is the concept of social time significant in accounting for how reciprocity in the care of dependent older people in receipt of formal care is promoted or impeded?
(3) In what ways is the concept of meaning making, as it relates to individuals and their spirituality, significant in accounting for how reciprocity in the care of dependent older people in receipt of formal care is promoted or impeded?
(4) In what ways is the concept of meaning making, as it relates to discourse and institutionalised patterns of power, significant in accounting for how reciprocity in the care of dependent older people in receipt of formal care is promoted or impeded?

Research Design

The design of the study has been to use interviewing as a method of data production and latent thematic analysis as an analytic tool to draw out aspects of commonality and difference in the participants' experiences of reciprocity, or its absence. In the discussion below, I explain my rationale for designing the study as I did, and reflect on how matters of reflexivity and academic and ethical credibility were recognised and addressed throughout the journey.

Data Collection

Why Interviews?

I considered that data production by individual interview was consistent with my aims and values in the following four ways:

1. It seemed to me to be the best way to try to get an insight into people's perceptions of their experience. That is, it fitted with the aim of my study, in that it would allow me some, albeit limited, access to their world-view. In trying to keep my work as a practitioner focused, when embarking on any project I have used the systematic practice framework outlined by Thompson (2009)—asking myself the questions (a) what am I trying to achieve? (b) how am I going to achieve it? and (c) how will I know when I have achieved it? When asking myself the same questions as a researcher it seemed that, if I was

seeking to explore people's perceptions then I had to hear their stories, so the use of in-depth interviews was consistent with that aim.

I am aware that, when asking people about their past, the data produced will be their memory or perception of the past, rather than the *actual* past which is supposed by some to have existed. As Becker (1992) comments:

> An unending sequence of "nows" feeds the flow of time. Indeed, we do remember the past and fantasize about the future, but we only do so here and now. The past that I remembered yesterday may be completely different from today's past; each past appears within the present moment and is influenced by it. The past, then, is always the present past. (p. 26)

However, I did not see that as a problem, given that perception and meaning making are the cornerstones of the methodological approach I have chosen and the focus of my research.

2. Interviewing participants also fitted with the phenomenological approach I have adopted and I have been reassured by Grbich (2007) that it is an appropriate way to get as close as possible to the 'essence' of the experience in question. This is not to use the term in the sense of a basic 'truth', but in the sense of a distillation of each person's particular experience, and its unique meaning to them. As a phenomenological researcher, I am interested in multiple meanings and unique ways of making sense of experience—hence the appropriateness of one-to-one interviews.

3. Furthermore, it also fitted with my professional commitment to facilitating the hearing of service user voices by practitioners and service providers (Braye 2000; Webb 2008; Hernandez et al. 2010). As an experienced interviewer in my professional life, I could see the potential for a well-designed interview schedule and a sensitively executed interview technique, to produce rich and informative insights that service providers (and policy makers) need to hear if power imbalances based on ageist assumptions about competence are to be addressed.

4. It fitted with my desire to address what I considered to be a lack of commitment to, and a dearth of opportunity for, reciprocal activity by recipients of care. As I discuss later in this chapter, I presented the call to participate as an opportunity to play a part in addressing concerns about care provision for older people which does not recognise and respect their personhood, rights and need for continuing growth. Reflexivity is not just an issue for me as a researcher, but something that is part of everyday living as people try to make sense of the world in which they operate (Ferguson 2001). The lives of older people are often characterised by a myriad of changes which need making sense of, especially in terms of losses due to experiences such as illness, bereavement and relocation; (Katz 2002; Lloyd 2002; Renzenbrink 2004) yet the impact of those changes in terms of identity may go unnoticed. The following excerpt from the work of Beck and Beck-Gernsheim (2001) highlights the dynamic between reflexivity and the multifaceted nature of identity.

Data Collection   75

> Something like individual distinctiveness really appears for the first time through the combination of social crises in which individuals are forced to think, act and live. It becomes normal to test out a number of different mixes; several overlapping identities are discovered and a life is constructed out of their combination. (p. 27)

While ageist discourses would have us believe that frail older people have only one identity (that based on being in need and receipt of care), interviewing them had the potential to highlight the fallacy of that assumption by reframing them as competent individuals who can give as well as receive.

When designing the study I gave some consideration to using focus groups (instead of individual interviews) to hear about people's experiences, but decided against it for fear of losing the depth of detail that I expected would be produced by one-to-one encounters. I wondered, too, whether the focus on individual perception might be distorted by the 'groupthink' effect (Janis 1982), which can occur in a group setting when individuals are influenced by the views of others, social mores and so on, to suppress their own views and concur with a supposedly universally held opinion.

Why an Indian Dimension?

Given my emphasis on:

- context, and how it contributes in general to meaning making; and
- the pivotal importance of the cultural level of shared meanings within the PCS framework that informs this study,

I considered myself fortunate to have the opportunity to interview dependent older people in an urban location in southern India, in addition to the sample I intended be drawn from urban or semi-urban locations in the UK. The decision to include Indian elders in the sample was largely based on opportunism, given that links with social work colleagues in Tamil Nadu, in southern India, had already been established by university colleagues, and I had already made contact with social work academics in Chennai as part of an international research initiative. Given the high profile of cultural and structural context in my study, I could see the potential for the voices of those experiencing reciprocity (or its absence) in a different context from the UK to enrich the study and so I took the opportunity gladly.

And so, while I had not originally set out to undertake a comparative study, but to explore experiences of reciprocity in one cultural setting, I considered that hearing the perspectives of people from a different cultural setting had the potential to highlight points of similarity and difference that could enrich my study, based as it is on the Gadamerian premise discussed earlier that knowledge is produced through a 'fusion of horizons'. While I do not consider it to be a comparative study, nevertheless, a *degree* of comparative element was therefore incorporated into my research design as the process developed, defensible in line with the qualitative paradigm's stance on flexibility in the face of unanticipated events (Lewis 2003) and consistent with my aim to have variety of perspective and context.

Data Analysis

Aiming to explore whether there are themes or shared meanings apparent in a collection of individual narratives, I chose latent thematic analysis as the most appropriate way of analysing the data produced, for reasons explained below. Having earlier been alerted to the existence of interpretative phenomenological analysis (IPA), I initially gave some consideration to using it as an analytical tool for this study. However, on further research I realised that, while it has the following emphases which are consistent with my methodology:

- an emphasis on sense making;
- an inductive focus; and
- a respect for the fact that service users are the experts on their own lives,

I concluded that the approach has too much of an individualistic focus for me to adopt it in this instance. While the social and cultural context is by no means ignored (Reid et al. 2005), there would appear to be a primary focus on understanding how individuals make sense of their experience, as is implicit in the following comment from Smith and Osborn (2003):

> IPA has a theoretical commitment to the person as a cognitive, linguistic, affective and physical being and assumes a chain of connection between people's talk and their thinking and emotional state…thus IPA and mainstream psychology converge in being interested in examining how people think about what is happening to them but diverge in deciding how this thinking can best be studied. (p. 52)

While I would agree with Reid et al. (2005) that IPA has the potential to contribute to biopsychosocial perspectives, I have nevertheless discounted it because I consider that IPA's focus on how individuals make sense of their experiences, and my own emphasis on shared meanings and ideologies, places us too far apart on the continuum between the individual and their social context for it to be a useful analytical tool for addressing the research questions I set myself.

Braun and Clarke (2006) argue that thematic analysis, while a method in its own right, can be used within a variety of analytical traditions, and manifests differently across different approaches. Broadly speaking, it is a method for identifying and analysing themes within data. As reported by Braun and Clarke (ibid.) it: 'can be an essentialist or realist method, which reports experiences, meanings, and the reality of participants, or it can be a constructionist method, which examines the ways in which events, realities, meanings, experiences and so on are the effects of a range of discourses operating within society' (p. 81). It is because my focus is on the power of discriminatory processes to operate largely unnoticed (Mullaly 2002; Thompson 2011) that the process of thematic analysis that I undertook can be said to be one of *latent* thematic analysis—that is, it was the underlying meaning making which informed the semantic content of the transcripts that was of particular interest.

Latent thematic analysis has provided the opportunity to explore the subtleties of responses in different cultural contexts—a point which Graneheim and Lundman (2004) make in its favour:

> thematic analysis conducted within a constructionist perspective cannot and does not seek to focus on motivation or individual psychologies, but instead seeks to theorize the socio-cultural contexts, and the structural conditions, that enable the individual accounts that are provided. (p. 85)

From the constructionist approach that I favour, latent thematic analysis appeared to, and indeed proved to be, flexible enough to allow me to pre-define broad themes (related to the four research questions which formed the basis of this exploratory study) as categories within which to recognise, organise and reflect on, further sub-themes. Latent thematic analysis structured around the four-part phenomenological framework already outlined therefore seemed an entirely appropriate way to go in terms of data analysis. Braun and Clarke (ibid) alert researchers who use thematic analysis to the potential pitfalls of;

(a) failing to progress beyond identifying themes *per se*; and
(b) not making explicit their bias in terms of interpreting the data produced.

Given that latent thematic analysis requires examination of the assumptions within the narratives and that, by pre-defining broad themes I had ensured that themes would emerge from the process already somewhat theorised, I consider that I had guarded against at least the first of these concerns. As for the second, I argue that it is neither possible nor useful to deny the 'researcher effect' and would agree with Fook's comment that revisiting transcripts at different times, and from different theoretical perspectives, is likely to produce different interpretations of the same material (Fook 2001). While my use of thematic analysis is explicitly theoretical, in that I approach the data with a pre-defined framework of concepts rather than searching the data in an inductive way, I have attempted to introduce an element of academic rigour by asking an experienced colleague to scan some of the interview transcripts to cross check whether her understanding of the emerging issues matched my own.

Reflections on Power

Given that this research is an anti-ageist endeavour, power is a concept that I could not allow myself to ignore when designing the study. I faced a number of dilemmas, some of which I could resolve and others that I had to accept as limitations to the study. I will return to limitations later in this chapter, and in the concluding one, but offer the following as examples of how reflection on my location as a researcher had implications for the research design:

- First, I was aware that my use of the term 'dependent', which I used as a criterion when recruiting the sample, carries the potential to reinforce the stereotype of old age as necessarily a time of dependency and vulnerability. Furthermore, it might seem that I was setting myself apart from those I interviewed by suggesting that a 'dependency' label is not appropriate to me, but only to them. It has long been recognised as a contested concept (Phillipson et al. 1986; Kennedy and Minkler 1998; Estes et al. 2003; Fine and Glendenning 2005) and I would agree that, given that we are all dependent on each other for a variety of reasons, and at different points in our lives, I am perhaps guilty of perpetuating an ageist stereotype in my use of a dependent/independent dichotomy rather than the continuum of interdependency that better describes everyone's life experience.

  I considered, then, how using the term 'dependent' in the interview schedule might affect the outcome and whether it, in itself, could constitute one of the issues that impede reciprocity by suggesting that the participants' status was pre-defined in line with ageist assumptions about competence. I was aware of how the way I worded the interview schedule, and conducted the interviews, would have at least some impact on the responses produced, however hard I worked at minimising that effect. I decided that the dependent/independent dichotomy was a necessary evil if participants were to understand how my questions would relate to temporal changes in care provision, and their perception of it, but remained alert that 'dependency' is a subjective and emotive term which might need qualifying or reframing, by way of prompts, at the interview stage.
- The decision to conduct interviews in Tamil Nadu also raised concerns for me about the power relationship between me as a researcher, and those people I planned to interview there. I had no real way of knowing how those participants would perceive me, but I could not discount the possibility that I might be seen as someone representative of a colonialist regime that they had either experienced personally, or heard accounts of, as being oppressive. When visiting Tamil Nadu to plan the study, I met a few older people on the streets who acted in a deferential manner, which suggested to me that my concerns were perhaps not without foundation. I decided that addressing, or at least minimising, this real or perceived power imbalance lay in presenting the study to potential participants as one of reciprocity and shared endeavour, rather than imposition.
- My third point relates to the dilemma I faced about whether or not to disclose my professional background to participants, in light of the effect that this knowledge would have on the dynamics of the research activity. In rejecting the assumption by the early phenomenologist Husserl, that it is necessary and possible to suspend or 'bracket' one's own experience when exploring phenomena (Moustakas 1994; Moran 2000), I am influenced by later phenomenologists such as Gadamer (2004) to claim that this is not something I am able to, or have ever intended to do. I consider myself to be part of the dynamic, accepting that my experience, rationale and values will necessarily impact on what I choose to research, why I choose to research it, how I interpret the results and what use I put them to—this all being part of what Nightingale and Cromby (1999) describe as epistemological reflexivity. Underpinned as it is by the social constructionist paradigm, I consider my

research journey to be one of constructing, rather than discovering, knowledge and, as discussed earlier, that knowledge is open to interpretation. Meaning making needs context and I would consider it an unfair relationship of power if I were aware of the context of the experiences of those I would be interviewing, but the participants were unaware of the context of mine.

I accept that it could be argued that knowing my background as a nurse, social worker or researcher had the potential to affect what was disclosed—especially given that it had been reported by one of the agencies I used as a conduit that some of the participants had been on the receiving end of poor quality care from individual social workers and social work agencies, and so could have been mistrustful of me as a consequence. However, in the spirit of partnership that I espouse, I chose to disclose my professional background in order to promote a good working relationship based on openness and trust, and to give credence to the interviewees' capacity to process and interpret information on their own terms—what Riach (2009) appears to be describing in the concept of 'participant-centred reflexivity'.

Denscombe (2003) advises researchers to consider the 'interviewer effect' carefully but he too refers to context and meaning making when he says that:

> The impact of the researcher's personal identity, of course, will depend on who is being interviewed. It is not, strictly speaking, the identity in its own right that affects the data, but what the researcher's identity means as far as the person being interviewed is concerned. (pp. 169–170)

As I considered all of those I interviewed to have mental capacity, and their relationship with social work, social care and health professionals to be part of the dynamic of their stories, I chose divulging my professional identity as my 'default setting', being prepared to consider withholding it only in what I considered to be exceptional circumstances. In the event, I never felt that it was necessary to do so.

Research Methods

Preparation for fieldwork

A key issue in research ethics has been that of vulnerability, and whether conducting research with a group of people considered to be vulnerable is ethically sound (Potter 2002). I consider that I countered this concern by framing the invitation to older people to take part in my study as an opportunity to work in partnership with me, rather than their being subjected to a process over which they have little or no control—that is, to be research partners rather than objects of research (Peace 2002). The promoting and facilitating of service user control has been central to my practice as a social worker, underpinned as it is by a professional value base that has rights and respect at its core (Banks 2006; Moss 2007). As such, I was mindful when planning this study that the researcher and care manager/therapist roles have distinct boundaries that I should respect. However, after reading what Barsky (2010) has to say about ethics, I wondered whether I was perhaps in danger of making those boundaries too rigid by not recognising the interplay between the two roles. I found Barsky's concept of 'the virtuous social work researcher' illuminating in that he maintains that, rather than 'policing' research by enforcing adherence to externally imposed ethical guidelines and codes of practice, it is social work 'virtues' (including what he refers to as caring, generosity, and concern for others) that lay the foundation for creating a culture of ethical research.

Revisiting my rationale in the light of his comments strengthened my conviction that my research is ethically sound because it is underpinned by the same ethical considerations that underpin my practice. What matters to me as a practitioner (such as addressing oppression, respecting dignity and rights, being transparent and honest in relationships and transactions) are the same issues that matter to me as a researcher.

I came to see the researcher role as an extension of the advocacy role I have often played, in that my guiding principle as a practitioner is that there is expertise not only in the practitioner and researcher voice, but in the service user voice too.

Before operationalising the research design, I had discussions with social work colleagues in Chennai to check whether commonly used terms relating to social welfare carried the same meaning in India as they did in the UK. It became apparent from these discussions that I could not take shared meanings for granted. For example, in India, the term 'social worker' appears not necessarily to carry the same connotation of professional status and training as it does in the UK and some of those I interviewed appeared to use it in a way more akin to 'charity worker' or 'volunteer'. I avoided the term 'care package' because it did not have currency in social care provision in Chennai at that time. Furthermore, on the advice of my Indian colleagues, I took care not to frame the participants' experiences as 'dependency' during the interviews, as they expected that it would be perceived as a form of insult.

Sample

As discussed in more detail later, the sample was purposive. That is, it was a non-random sample drawn from two urban/semi-urban locations chosen to reflect experience of reciprocity, or its absence, in contexts where formally delivered care provision existed. It was purposive in the sense that specialist knowledge was drawn on to select participants who would represent the population and experiences I intended to study. The relative paucity of care services in rural India suggested that it might be difficult to find people there who fitted the specified criteria of being in receipt of care beyond that provided by family. I was aware that while India is undergoing a period of rapid social change (Varma 2005; Farndon 2007), there is still a very significant divide between urban and rural living there. Hence, I confined the sample to urban dwellers.

The populations of both India and the UK are diverse and, in terms of social policy, there is a degree of variation within each country, as well as between them. For example, as I have already commented on, while there is a national plan for eldercare reform in India (National Policy for Older Persons, 1999) there is a disparity of provision at state level (Datta 2008), and in the UK there are policy differences since devolution. As some of the people I interviewed in the UK resided in North East Wales and some in England, this was something that I had to keep in mind if I were going to draw comparisons. However, the samples were not intended to be representative of their respective nations and I do not make that claim. As I was looking for aspects of commonality of experience, I decided that to draw the sample from urban populations in both countries would remove a potentially complicating variable from the equation.

I wanted to ensure that the participants were not disadvantaged by being unaware of the wider context of the study and therefore ensured that, when recruited to the sample, they were informed of the purpose and potential use of the findings so that they might perceive themselves as working alongside me in order to make a difference rather than as 'objects' of research (see information sheet—Appendix 2).

Given my commitment to the empowerment of older people, uppermost in my mind when planning the collection of data, was how best to ensure that participants would perceive the interview process as a positive experience rather than one in which they felt inconvenienced or, at worst, abused. This issue manifested as an ethical concern when it became clear that a history of some poorly conducted research in the field in Tamil Nadu was having an impact on people's willingness (or rather that of organisations who worked with older people) to participate in my project. When approaching agencies there with a view to constructing a sample, it became clear that, despite reassurances to them from an Indian social work colleague that my research was value driven and academically rigorous, the lack of those qualities in some of the research carried out in the field of eldercare in that region in recent years had contributed to putting research in a negative light.

Several administrators of residential establishments made the point that being interviewed was potentially stressful for participants and that, as they saw very little in the way of dissemination of results which would make that stress worthwhile, they felt justified in putting blanket bans on any research involving those they considered to be vulnerable. While this did present me with some problems, I had to agree that research which involved an abuse of power *should* be highlighted as unethical, and that their attempt to protect their clients was laudable, even though I considered a ban on all research to be misguided. Given this experience in the early stages of trying to construct a sample, I revisited my rationale and methodology but came to the conclusion that I was already doing everything I could to present my study as an ethically sound and worthwhile project which would give those taking part the opportunity to contribute to a knowledge base that might inform and improve welfare practice.

While considering how I could best counter any potential for a power differential to be detrimental to already vulnerable people, I was aware that I would be conducting research in two very different cultural environments, and that differing expectations of the role and status of older people, and of research activity itself, would need to be considered. Having taken advice from a leading academic at Stella Maris College, Chennai, I was reassured that the approval I had already obtained from a UK-based university ethics committee would be acceptable to those who might support my work in India. In the spirit of partnership to which I have already referred, I ensured that the managers of the residential establishments I approached when recruiting the sample, and indeed the participants themselves, were aware that I had sought, and received, ethical approval for the study before asking them specifically for permission to interview them, and to record and transcribe what they had to say about their experiences.

The specific issue of consent presented as an ethical concern at one point, in that, while in the UK, the issue of consent was in all cases considered to reside with the individuals I approached, the founder and manager of one of the homes in Chennai insisted that there was no need to seek individual permission from potential interviewees, as she claimed the authority to grant it on behalf of those living under her care. While she did not appear to see this as an abuse of power, I most certainly did and, regardless of her 'instructions', obtained written consent from each individual participant before conducting the interviews. In discussing agency and ethics, Nespor and Groenke (2009) make the point that:

> a key component of the researcher's obligation is to accurately explain to potential participants what is being asked of them, to lay out the risks and benefits of participation, and then to allow people full exercise of their agency in giving or refusing informed consent. (p. 997)

I consider the key word here to be 'accurately' and, therefore, insisted on taking control of the process of acquiring consent in order to ensure that my intentions and values as a researcher would be accurately transmitted. By doing so, I could feel confident that the potential for any misunderstanding, or misrepresentation, of them on the message's journey from me to those people I would be interviewing would be minimised. I avoided the use of professional jargon in both written and spoken exchanges to this end, especially at the point of obtaining consent, as this was my first point of contact—a highly significant one in terms of establishing relationships of trust and respect (Hamer 2006).

Inclusion and Exclusion Criteria

In order to ensure that those I interviewed would have had experience of the phenomenon in question, the sample had to be a purposive one. For that reason, I stipulated the following criteria when searching for potential research participants:

- That they be aged 70 years or over. A lower age limit of 70 years was specified, although this was to some extent a fairly arbitrary one, given that there is no absolute consensus on what constitutes old age (Phillipson 1982; Bytheway 2011). Since my aim was to explore what impedes or promotes reciprocity in the care of dependent older people I had to ensure that, as far as possible, the sample was drawn from a population that would generally be conceptualised as 'old', and in which I would be likely to also find people who had experienced a significant level of dependency on formal care support.

  In terms of age as a criterion, I had to consider differences in life expectancy as, although it is static in neither country, average life expectancy in India is lower

Research Methods   83

than in the United Kingdom (Chakraborti 2004; Census of India 2009). I had specified a lower age limit of 70 years for those older people in the UK, so that there would be the likelihood of finding dependent older people in that age category. In order to have the lower age specification better reflect the demography of southern India, and maximise the chance of finding individuals who fitted the criteria outlined, I lowered it to 60 years for the Indian sample. In the event I need not have worried—men and women over 70 were not hard to find and my first interviewee was, at 109 years old, significantly older than the oldest member of the UK sample.

- That they had acquired their significant level of dependency after entering the phase of life generally considered to be old age, rather than having aged with a lifelong disability or debilitating condition. The intention here was to minimise the potential for issues connected with ageism, and age-related discrimination, to be obscured by those related more specifically to disability.
- That they resided in urban or semi-urban locations. There is evidence to suggest that there are disparities between the experiences of urban and rural dwellers in both India and the UK (but particularly in India), in terms of life opportunities and welfare provision (Farndon 2007; Cheers and Pugh 2010). I therefore chose to exclude rural dwellers in both samples to minimise the effects of the rural–urban divide as a variable.
- That they not be suffering from dementia, or any condition or significant learning disability that could affect memory or recall. I chose to exclude people with such conditions from the sample because I needed participants to be able to recall their memories of opportunities for reciprocity earlier in their lives.

Size. The sample size was deliberately small in both countries, as my aim was to explore perceptions in some depth—producing a small amount of data that would help illuminate people's experience of the opportunity to reciprocate, or its absence, and add to the knowledge base of eldercare—rather than generating material that would provide breadth of coverage and allow for generalisability. As Armour et al. (2009) point out, where phenomenological approaches are employed there is little point in generating large samples because: 'the focus is on in-depth study of information-rich cases and the extraction of essential characteristics of a phenomenon, rather than on variation between research respondents (p. 106).

In order to keep the data produced to manageable proportions in terms of time and cost, I had anticipated interviewing between 6 and 10 people in each location. Because of restrictions relating to available time and access to institutions, I was able to interview seven people in Tamil Nadu. This was in no way disappointing as a sample because the data produced was rich in detail, and informal analysis immediately after the event was indicating that a number of significant themes were already emerging from the data. On my return to the UK I decided to construct a similar-sized sample, and to limit it to seven people unless I felt that such a sample was failing to indicate the emergence of themes that were informed by my research questions. In the event, the UK sample also proved to be rich in informative data and I did not exceed the revised sample of 14 people in total.

Gender. I intended that both samples include males and females, in order that issues around gender not be excluded from the findings. Research has indicated that there are different expectations of, and from, men and women who provide or receive care (Rose and Bruce 1995; Calasanti 2003; Blood 2010). I guessed that this might have implications for expectations around reciprocity, especially given the potential for the 'commodification' of care (Ungerson 2000) to impact on whether, in contexts where care is a bought commodity rather than a friendship or kinship obligation, women in particular feel obligated to reciprocate for care received in a way that men might not if they have been brought up in, or maintained, patriarchal households. Furthermore, including men and women in the Indian sample opened up the potential for an extra layer of richness in the data.

Religious background. I did not specify any particular religious background in the hope that a variety of world-views would be represented. In the event, all participants in the India sample came from the Hindu faith, although different class and caste affiliations were represented within that classification. Jacobs (2010) describes how class and caste co-exist in urban India:

> Class as a social category has largely been overlooked, primarily because scholars have typically seen it as an alien concept that cannot be imposed on the Indian context. However, many Indians do identify themselves as being middle class. The middle class in India is not a clearly defined category, but can be understood as primarily being an urban phenomenon where meritocracy is regarded as taking precedence over, although not totally replacing, ascriptive social categories. (p. 34)

As they have implications for social expectations around reciprocity and the ability to support those felt to be less fortunate than themselves, this had the potential to enrich the results. As I discuss later, in Chap. 6, within the context of 'Hindu-ness' there are expectations relating to reciprocity, and these provide a vein of discussion about what impedes or promotes it at the levels of both culture and ideology. While other religious perspectives may have provided additional layers of meaning to the discussion, and would indeed make an interesting basis for further studies, I did not see the absence of representatives from non-Hindu perspectives in that sample as problematic for this study.

Type of care provision received. While I had specified that participants be in receipt of care packages, or living in residential care, I did not specify that both be represented in each sample. During my preliminary research into the feasibility of this study, I had been led to believe that both residential and domiciliary care agencies existed in Chennai, as is the case in the UK. However, while visiting the city with the intention of constructing a sample, I realised that, although the term 'domiciliary care' is used in both India and the UK, there is a significant difference in terms of the work undertaken by agencies in both contexts.

Where such agencies existed in Chennai, their focus was on providing support (most often staple foodstuffs) to families rather than individuals, particularly those in deprived areas or circumstances. While, technically speaking, some dependent older people within those families could be said to have been in receipt of help from domiciliary care agencies, supporting them was not the primary aim of those agencies, which were only very peripherally involved in eldercare. In the UK, individuals fulfilling the criterion of being in receipt of formal care were to be found in both domiciliary and residential settings, but it seemed that, in Chennai at the time of the study, formal care equated with residential care and so the Indian sample was drawn only from those living in residential care facilities.

In the process of constructing the sample, I was reminded of the extent to which old age is socially constructed (Phillipson 1982; Wilson 2000) and of how differences of conceptualisation inform what older people could expect in the way of support in the two welfare contexts. For example, I found myself considering why I had imagined that a society in which the expectation for families to provide care for their elderly relatives remained a dominant norm would *need* domiciliary agencies. Although I deviated from my original plan to recruit the sample from people representing those in receipt of support in both settings in both countries, I maintain that drawing the sample from a population in residential care only in India, but from a residential and domiciliary care population in the UK, reflected the prevailing welfare provision in each context and was therefore justifiable.

Recruitment. As discussed, one of the issues I considered before engaging in constructing a sample in both countries was my awareness of how research involving vulnerable, and often marginalised, individuals is fraught with ethical issues, including the potential for the research process to cause harm in some way, or transgress basic rights, such as self-determination. My decision to use an opportunistic method of sampling was informed by a desire to have people *choose* to participate—that is, to opt in rather than being presented with the *fait accompli* of having been chosen as part of a sample from which they would have had to opt out if they chose not to participate, as would have been the case in a random sample. While allowing someone to withdraw from a predetermined list might have ticked an ethical box, I would argue that this is much more difficult to do than responding to an offer to participate, even where no external pressure is applied.

Although aware that self-selection has the potential to produce bias, in that there is a danger that those with an 'extreme' view are more likely to respond to an invitation to participate, I chose to draw both samples from organisations who expressed a willingness to offer my invitation to participate to their clients. The possibility of bias was minimised by asking those representing each organisation not to go into any more detail about the focus of the study than I had done so myself in the introductory letter (see Appendix 2). By doing this I hoped to optimise the chance that people would agree to participate because of an altruistic desire to help others in a general sense, rather than because of having a particular axe to grind, and seeing these interviews as an opportunity to grind it. Despite the concerns expressed by (Holland 2005) that;

> The expectations of altruistic involvement is wearing a bit thin though for some older people today—including people in the groups that feel themselves to be over-researched; people who are very busy; people who are very private—and those just not feeling very altruistic at the time (p. 9),

I hoped that, by using care agency representatives as 'conduits', I could build up a sample of people who, through their willingness to share their experiences with me, *did* have an altruistic motivation similar my own, and could be considered partners in a shared research endeavour, rather than as people *being* researched. Informal feedback after the event from several of the interviewees in both India and the UK, and from those managing the organisations, suggested that this was how the process had been experienced from their perspective, which was encouraging.

Pilot Study

I created the opportunity to test out my assumption that using my interview schedule would produce useable and relevant data by using it in a pilot study to test out whether:

- the interview schedule was long enough to produce useful data, but not so long that it would prove too tiring or onerous an experience for people who, by definition, would be experiencing compromised health or ability;
- the questions posed would resonate with people's experiences and engage them to the extent that they would want to share their experiences with me;
- given that loss experiences would feature in the schedule of questions, the strategy of ending the interview with questions that related to positive, rather than negative, outcomes would minimise the potential for distress to be caused but not noticed and addressed.

Two people, a man in his 80s and a woman in her 90s (both confined to their homes in the UK because of debilitating conditions associated with strokes and falls), agreed to be interviewed. They were informed of its nature as a pilot study, its purpose and that the transcripts would not be made available to anyone but them (if they chose to have a copy) and me. Both declined to have their own transcripts of the interviews. At first this perturbed me because it challenged my consideration of them as research partners (Ross et al. 2005; Dewar 2005) who had the right to scrutinise what I was attributing to them but, on reflection, I came to realise that I was basing my assumption that they should have that information (and, indeed would want it) from my own perspective as a researcher needing to feel satisfied that I was 'doing right' by them. From discussions prior to, and during, the pilot interviews, it had become clear to me that both participants saw the process as an act of altruism—a way of helping their peers by way of helping me. From that perspective, they considered their part in the process to be over once the interview had finished, and saw no need to have a record of it.

In that situation, my understanding of the participants' perspective of their role, the limits they imposed on it and my action of respecting their decision to challenge what I initially considered to be necessary research protocol, was informed by what Barsky (2010) refers to as phronesis, or practice wisdom:

> For SWRs [social work researchers] phronesis has implications for research design and implementation. SWRs do not simply rely on textbook information and research protocols for how to design and implement research. They make use of their experience working with clients and research participants to determine how to act in particular situations. (p. 5)

The pilot study proved to be a useful exercise in that the data produced validated my choice to use interviewing as a methodological tool, and highlighted only one significant issue that needed addressing before using the schedule for the main study. In the original schedule of questions used in the pilot study, when I asked people to focus on the period before they had become significantly dependent on others, and to consider what knowledge or skills they felt they had at that point, it became clear that this was a difficult question for them to answer without a fairly lengthy time allowed for reflection, and the use of several prompts. This may not have been entirely connected with powers of recall, but perhaps also had something to do with a reluctance to 'blow their own trumpet'. Indeed, it may also have been because self-reflection to that degree is not a skill many people have, given that it does not often figure in most people's day to day lives. As Elliott (2005) remarks:

> the meanings and understandings that individuals attach to their experiences are not necessarily pre-formed and available for collection, rather the task of making sense of experiences will be an intrinsic part of the research process. (p. 24)

Testing out the interview schedule reminded me that I was asking 'big' questions for which I could not expect quick or unconsidered responses. For example, the male participant who was the first to be interviewed in the pilot study seemed reluctant to talk about the skills he had possessed in his early life as a head teacher. It transpired that this was not because he was reluctant to talk about them, but rather that he did not consider himself to have been skilled. This led me to consider how I could facilitate the insights I was seeking without asking leading questions that would introduce anything more than the minimum of bias that is necessarily part of the 'subjective lens' of the qualitative researcher (Thyer 2001). Adding prompts that incorporated 'usefulness', 'what you did' or 'what you could offer that made you feel proud or good about yourself' broadened the concept out from the original one of skills and led to a less laboured and more fruitful response from the second participant.

The process also prompted me to record, in subsequent interviews, albeit in brief note form so as not to disturb the flow of the interview, the skills or attributes recalled from the earlier phase of the life course before moving on to the second phase of the schedule. This refinement of the interview process facilitated an easier linking of the recollection part and the current experience part (that is, before and after becoming dependent) and was employed in the 14 interviews that followed.

After these changes, and some minor amendments to the wording of prompts and the ordering of questions, I felt confident that my data collection tool was fit for purpose.

Interviews

In terms of interviewing people in Tamil Nadu, I was aware that English was fairly widely spoken, especially in professional and business communities where it is often the primary medium of communication. While some interviewees appeared to have an excellent grasp of English, I was mindful that to expect someone to engage with matters of a personal nature (and on an emotional level) in a medium other than their first language could constitute an unfair use of power on my part (Davies 2009; Roberts et al. 2011) and so only agreed to conduct the interviews in English if I was asked to do so. However, even where the language was supposedly a shared one, the potential for miscommunication remained. As becomes evident later, as my inquiry progressed I became aware of some quite significant differences between English as spoken in India, and that spoken in the UK (John 2007) and had to remain alert to the ethnocentrism which could have featured in my work had I assumed that participants were understanding our dialogues in the same way that I was.

I had anticipated needing the services of a translator for instances where the interviewee spoke one of the Indian languages or dialects, but the person chosen also proved to be adept at noticing and pointing out instances where there was the potential for misinterpretation when English was the medium used. As Sunderland (2004) and Ryen (2008) point out, the context in which a language is spoken is significant, in that the transmission of information is not neutral. In anticipation of the danger of interpreting comments from only my own perspective I had sought the services of someone with good observational as well as translation skills, and who could understand my perspective to some extent, while also remaining outside of it. I chose to use the services of a student about to graduate with a Master's degree in social work studies, recommended because of her proven communication and engagement skills.

After our initial meeting I felt as confident as I could be that the process she was undertaking on my behalf was underpinned by our common values and skills as social workers. The early confidence in those values and skills was borne out as I observed her cross-checking that interviewees understood my questions, and that my understanding of their responses matched with hers, and related appropriately to the questions. She appeared to share my regard for body language as an integral part of the process of communication and it was clear from the recommendations that preceded her, and the practice I observed in the field, that she understood the nuances of different dialects. Furthermore, she could highlight when this had the potential to be a point of misunderstanding, such as when she had to, on one occasion, clarify which phase of the life course I was asking about.

Because the research questions related to temporality and meaning making, these concepts informed the interview schedule's 'before and after dependency' format (see Appendix 1) which asked participants to reflect on how they had reciprocated earlier in their lives, and the extent to which that identity was being maintained or undermined in old age.

I considered conducting two interviews which related to the two life stages but decided against it, partly for reasons of practicality and cost, but also because I was concerned that the relationship between the sense of identity felt in the 'before and after onset of old age' scenarios might have been lost by separating out the two stages. As can be seen from the copy of the interview schedule, I designed six questions and a prompts list, with a view to maximising the chance of eliciting information which would address the research questions—that is, not just about their own perspectives on reciprocity, but also their views on whether others involved in providing or organising their care were aware of, and respected them.

My respect for confidentiality alerted me to consider where the interviews should be conducted, so as to safeguard rights to privacy and protection from exploitation. Conducting the UK interviews in private was not generally a problem, but it was difficult in the Indian context to ensure that participants had the opportunity to give an individual and unique response—private space was at a premium and interruptions were frequent. However, the focus on confidential and individual responses was maintained throughout, despite the difficulties.

The necessity of recording the interviews for subsequent transcription and analysis was a matter for some concern, as I thought it might 'formalise' the proceedings and encourage participants to behave or respond in ways other than they normally would. In making my status as a researcher more explicit, and reminding them that this was an unusual situation, I was concerned that it might prompt them to say what they thought I wanted to hear. Riach (2009) reminds us of Bourdieu's concern that research interactions can lose layers of richness and complexity if they become too instrumental. Hall (2001) recalls similar concerns:

> My feelings about qualitative interviewing as a research technique are mixed. Interviews can be extremely useful and revealing, but the tape-recorder has a tendency to elicit particular sorts of narrative from respondents – those which respondents feel are appropriate to the interview context. (p. 55)

In order to 'neutralise' the dynamics as far as possible, having shown the participants the recording equipment, and re-confirmed that they had given permission for me to record, transcribe and store their responses, it was placed in the least obtrusive place that would allow for a good recording. In most cases, it appeared to get forgotten about by everyone except me who had the responsibility of ensuring that neither recording space nor battery had run out!

Interview times ranged from around 40 min to almost 2 hrs, according to how much the participants wanted to talk and were able to tolerate the process. Where I felt them to be tiring, or they had indicated that they were nearing the end of their capacity or willingness to talk, I truncated the schedule to allow for each question to be addressed in some way, however small, by each participant. No interviews were therefore incomplete.

On reviewing the interview schedule prior to the commencement of the data collection phase, I had considered the strategies I might need to use in order to minimise the potential of falling into the trap of interviewing as a care manager/therapist working to promote change, rather than as a researcher. After all, it had been my observations in those roles that had made me keen to undertake research that might inform practice and promote change. As Armour et al. (2009) comment, having some knowledge of the phenomenon under scrutiny can be problematic:

> These efforts to expand knowledge raise important questions. For example, it is commonly understood that qualitative researchers must strike a balance between knowing enough to pose and probe the research question and knowing too much. The latter increases the

likelihood that researchers will make assumptions about phenomenon because pre-shaped knowledge reduces their ability to maintain an open and discovering position. (p. 116)

Yet, as others have argued (Legard et al. 2003; Thyer 2001; Roulston 2010) it is neither possible nor desirable to produce or analyse data other without it being mediated through the prisms of researcher values and experience. As Roulston (2010) comments:

> Given that in qualitative interviews the researcher is the instrument, there is no escape from the self. Whether acknowledged or not, researcher selves are implicated in every aspect of a research project - from the formulation and design of a study, to the interview interaction, and analysis and representation of interview data. (p. 127)

Reminding myself that the rationale for my study was to facilitate the hearing of service users' 'horizons', or perspective, helped me to take a step back and portray myself as a facilitator, rather than a director, of the interview process.

Transcription

I chose to transcribe the interviews I recorded in India myself because my familiarity with the setting, and body language and accents used, reduced the potential for contextual cues to be lost in the process. While listening to the translator in the field, I was careful to watch the participants and make notes at that stage, on which I could reflect later when transcribing and analysing. The process of transcription by a third party has the potential to make the finished script even more removed from the original dialogue by introducing the possibility of an extra layer of interpretation in the act of representing someone's voice (Roulston 2010). To guard against this, I ensured that the UK interviews were transcribed by a very competent and experienced audio typist with whom I had worked before, and knew to be adept at recording non-verbal cues such as pauses, and who I could trust not to guess when unsure of a word or phrase used. I do not underestimate the role she played in the research process. Her prompt and accurate work enabled me to add my own notes about context, such as group dynamics and cultural expectations, to produce the most reliable recording of events that could be managed.

In a personal message, this typist reported to me that having access to my schedule of questions had caused her to reflect on:

(a) her elderly mother's comments about 'being on the scrapheap of life', and how she might boost her mother's very low self-esteem by talking to her about the considerable ways in which she continues to give to others, such as teaching her great-granddaughter to read; and
(b) how she herself has skills and knowledge that she keeps to herself because they make her feel competent, but fears losing her 'usefulness' in those respects as she gets older herself.

Receiving those insights from her was an unintended, but very welcome, consequence of what I had expected to be a purely instrumental relationship, and, though she had not been a research participant herself, nevertheless her experience chimes well with the earlier discussion of the transformational potential of research. She thanked me for the insight I had given her and I thanked her for the insight she had given me—reciprocity in action!

Analytical Tool

This was adapted to my purposes from the process of latent thematic analysis described by Braun and Clarke (2006).

Stage 1

I began the process of analysis by reading all 14 transcripts twice over, to remind myself of the narratives, and to immerse myself in the data. On a third reading, I began to sensitise myself to possible emerging patterns.

Stage 2

All 14 transcripts were annotated by me in respect of the emerging concepts, and an experienced researcher colleague was asked to undertake the same process independently on two of the transcripts, chosen at random. When a comparison was made between the annotated sample transcripts, and my own versions, some discrepancies became apparent. However, on the few occasions where concepts had been annotated differently, the differences related to linguistic terminology rather than conceptual differences, and so I felt confident that, while researcher bias could not have been eliminated entirely from this process, I had minimised it as far as reasonably possible.

Stage 3

Each identified concept was allocated, as appropriate, to one of four sections that related to the four-part phenomenological framework. Through a process of refinement, 19 themes emerged. While I considered them all to have significance by virtue of their having been raised by the participants at all, I paid some attention to quantifying that significance, as per the final stage below.

Stage 4

Across the whole set of transcripts:
- themes mentioned fewer than 20 times were categorised as 'significant';
- themes mentioned more than 20 times but fewer than 50 were categorised as 'very significant'; and
- themes mentioned more than 50 times were categorised as 'particularly significant'.

These 19 themes inform the presentation, in the next chapter, of the findings of this study. Before moving on to do so, I first recognise and address some of the methodological limitations, in anticipation of any criticism my approach may attract.

Methodological Limitations

As I have indicated throughout this chapter, I realise that the approach I have adopted has its limitations, as does any other approach. I have addressed some of those that relate to broader phenomenological issues in the previous chapter but, in terms of this study, I recognise the following limitations:

- The sample I used cannot be claimed as representative of all older people in receipt of formal care in the UK or southern India, but it was never intended to be.
- By using a purposive sample, I have to accept that I leave myself open to the accusation that I have chosen a sample which is bound to reflect the findings I am looking for. I would defend my methodology by saying that, to extend the knowledge base by exploring a phenomenon in order to understand it better, the sample *has* to comprise people who have experience of that phenomenon.
- Similarly, using pre-defined concepts as a framework within which to analyse data could be challenged by those who favour a purely inductive approach, because it could be argued that I might be influenced by my 'hunches' about reciprocity in eldercare, if not to manipulate data, then to unintentionally interpret it in a way which suits my focus. However, having worked and studied in the field for many years, my 'hunches' are what prompted this research and, as such, are necessarily part of the dynamic of the study. As Braun and Clarke (2006) maintain, engaging with the literature prior to thematic analysis is entirely appropriate for a theoretical thematic analysis such as mine.
- Interviewing participants once only proved to be limiting in terms of what extra insights might have emerged had participants had more time to reflect on my questions. Time and cost constraints, and recognition that the interview schedule had the potential to be physically, intellectually and emotionally demanding of the participants, prevented me from fully exploring some of the potentially significant issues raised. However, this turned out not to be a significant limitation as rich data was produced anyway.
- While participants may have disclosed other experiences or insights to another researcher, at a different time, or through a different mode of enquiry, I decided against a mixed methods study and resisted the call to triangulate data because my aim has been merely to check out whether my own observations of the significance of reciprocity in older people's lives is reflected in their experiences. As an exploratory study, if it does no more than highlight commonalities of experience relating to reciprocity in the contexts I am describing, then I will have succeeded in my aims of keeping it on the eldercare research agenda and highlighting the benefits of phenomenologically informed research.

Conclusion

To sum up, given my phenomenological focus on meaning making, and my intention to theorise the findings that emerge from the narratives of those I interviewed, I consider the methodological design, tool and methods employed to have been appropriate for purpose, and to have evidenced academic rigour. More specifically:

- interviewing on a one-to-one basis provided the 'insider' perspective I was seeking, and facilitated the intense listening that enabled me to better hear what the participants' narratives were reflecting beyond the immediate impression (Tehan 2007; Randall and McKim 2008);
- analysing the findings with reference to a framework of phenomenologically significant concepts facilitated the theorising of the findings, so that they were not just a collection of stories;
- leaving an audit trail has demonstrated transparency of intent and values (see appendices for sight of interview schedule, and so on);
- corroborating my perception of the concepts arising from the findings provides further evidence of academic rigour; and
- in addressing methodological limitations specifically, and making my values and concerns explicit throughout this chapter, and indeed the whole study, I lay claim to practising as a reflexive researcher.

I move now, in Chap. 5, to present the rich and enlightening findings that emerged.

References

Antonucci, T., & Jackson, J. (1989). Successful ageing and lifecourse reciprocity. In A. Warnes (Eds.), *Connecting gender and ageing: Changing roles and relationships*. Buckingham: Open University Press.

Arber, S., Davidson, K., & Ginn, J. (Eds.), (2003). *Gender and ageing: Changing roles and relationships*. Maidenhead: Open University Press.

Armour, M., Riveaux, S. L., & Bell, H. (2009). Using context to build rigor: Application to two hermeneutic phenomenological studies. *Qualitative Social Work, 8*(1), 101–122.

Attig, T. (2011). *How we grieve: Relearning the world* (revised ed.). New York: Oxford University Press.

Banks, S. (2006). *Ethics and values in social work* (3rd ed.). Basingstoke: Palgrave Macmillan.

Barsky, A. E. (2010). The virtuous social worker researcher. *Journal of Social Work Values and Ethics, 7*(1), 1–10.

Beck, U., & Beck-Gernsheim, E. (2001). *Individualization*. London: Sage.

Becker, C. S. (1992). *Living and relating: An introduction to phenomenology*. London: Sage.

Blood, I. (2010). *Older people with high support needs: how can we empower them to enjoy a better life'*. York: Joseph Rowntree Foundation.

Bowers, H., Mordey, M., Runnicles, D., Barker, S., Thomas, N., Wilkins, A., et al. (2011). *Not a one-way street: Research into older people's experiences of support based on mutuality and reciprocity*. York: Joseph Rowntree Foundation.

Braun, V., & Clarke, V. (2006). Using thematic analysis in psychology. *Qualitative Research in Psychology, 3*, 77–101.

Braye, S. (2000). Participation and involvement in social care an overview. In H. Kemshall & R. Littlechild (Eds.), *User involvement and participation in social care: Research informing practice*. London: Jessica Kingsley Publishers.

Bubek, D. (2000). Feminism in political philosophy: Women's difference. In Fricker and Hornsby (2000).

Bytheway, B. (2011). *Unmasking age: The significance of age for social research*. Bristol: The Policy Press.

Calasanti, T. (2003). Masculinities and care work in old age. In S. Arber et al. (Eds.), *Gender and aging: New directions*. New York: Routledge.

Census of India (2009). Government of India: Ministry of Home Affairs—www.censusindia.gov.in.

Chakraborti, R. D. (2004). *The greying of India: Population ageing in the context of Asia*. New Delhi: Sage.

Chasteen, A. L., Schwarz, N., & Park, D. C. (2002). The activation of ageing stereotypes in younger and older adults. *Journal of Gerontology: Psychological Sciences, 57B*, 540–547.

Chatterjee, D. P., Patnaik, P., & Chariar, V. M. (Eds.), (2008). *Discourses on ageing and dying*. London: Sage.

Cheers, B., & Pugh, R. (2010). *Rural social work: International perspectives*. Bristol: Policy Press.

Datta, A. (2008). Socio-ethical issues in the existing paradigm of care for the older persons: Emerging challenges and possible responses. In Chatterjee et al. (2008).

Davies, E. (2009). *Different words, different worlds?: The concept of language choice in social work and social care*. Cardiff: The Care Council for Wales.

de Beauvoir, S. (1977). *Old Age*. Harmondsworth: Penguin.

Denscombe, M. (2003). *The good research guide for small-scale social research projects* (2nd ed.). Maidenhead: Open University Press.

Dewar, B. J. (2005). Beyond tokenism: Involvement of older people in research—a framework for future development and understanding. *Journal of Clinical Nursing, 14*(3a), 48–53.

Elliott, J. (2005). *Using narrative in social research: Qualitative and quantitative approaches*. London: Sage.

Estes, C., Biggs, S., & Philipson, C. (2003). *Social theory, social policy and ageing*. Buckingham: Open University Press.

Farndon, J. (2007). *India booms: The breathtaking development and influence of modern India*. London: Virgin Books.

Ferguson, H. (2001). Social work, individualization and life politics. *British Journal of Social Work, 31*, 41–55.

Fine, M., & Glendenning, C. (2005). Dependence, independence of inter-dependence?: Revisiting the concepts of "care" and "dependency". *Ageing and Society, 25*(4), 601–621.

Fook, J. (2001). Identifying expert social work: Qualitative practitioner research. In I. Shaw & N. Gould (Eds.), *Qualitative research in social work*. London: Sage.

Fricker, M., & Hornsby, J. (Eds.), (2000). *The Cambridge companion to feminism in philosophy*. Cambridge: Cambridge University Press.

Fromm, E. (1955). *The Sane Society*. New York: Rinehart.

Gadamer, H.-G. (2004). *Truth and method* (3rd ed.). London: Continuum.

Giddens, A. (2009). *Sociology* (6th ed.). Cambridge: Polity Press.

Graham, H. (2007). *Unequal lives: Health and socioeconomic inequalities*. Maidenhead: Open University Press.

Graneheim, U. H., & Lundman, B. (2004). Qualitative content analysis in nursing research: Concepts, procedures and measures to achieve trustworthiness. *Nurse Education Today, 24*(2), 105–112.

Grbich, (2007). *Qualitative data analysis: An introduction*. London: Sage.

Hall, T. (2001). Caught not taught: Ethnographic research at a young people's accommodation project. In I. Shaw & N. Gould (Eds.), *Qualitative research in social work*. London: Sage.

References

Hamer, M. (2006). *The barefoot helper*. Lyme Regis: Russell House Publishing.
Harrington Meyer, M. (2000). *Care work: Gender*. London: Class and the Welfare State.
Hernandez, L., Robson, P., & Sampson, A. (2010). Towards integrated participants: Involving seldom heard users of social care services. *British Journal of Social Work, 4*, 714–736.
Holland, C. (2005). Some issues in recruiting and sampling from the older population. In Holland (2005).
Holland, C. (Ed.), (2005b). *Recruitment and sampling: Qualitative research with older people*. London: Centre for Policy on Ageing.
Jacobs, S. (2010). *Hinduism today*. London: Continuum.
Jamieson, A., & Victor, C. R. (Eds.), (2002) *Researching ageing and later life*. Buckingham: Open University Press.
Janis, I. (1982). *Groupthink: Psychological studies of policy decisions and fiascos* (2nd ed.). Boston: Houghton Mifflin.
John, B. K. (2007). *Entry from backside only: Hazaar Fundas of Indian-english*. India: New Delhi.
Jordan, B. (2008). *Welfare and well-being: Social value in public policy*. Bristol: The Policy Press.
Katz, J. (2002). Ill-health. In N. Thompson (Ed.), *Loss and grief: A guide for human services practitioners*. Basingstoke: Palgrave.
Kemshall, H., & Littlechild, R. (Eds.), (2000). *User involvement and participation in social care: Research informing practice*. London: Jessica Kingsley.
Kennedy, J., & Minkler, M. (1998). Disability theory and later life: Implications for critical gerontology. *International Journal of Health Services, 28*(4), 757–766.
Legard, R., Keegan, J., & Ward, K. (2003). In-depth interviews. In J. Ritchie & J. Lewis (Eds.), *Qualitative research practice—a guide for social science students and researchers*. London: Sage Publications.
Lewis, J. (2003). Design issues. In J. Ritchie & J. Lewis (Eds.), *Qualitative research. Practice*. London: Sage Publications.
Lloyd, M. (2002). A framework for working with loss. In N. Thompson (Ed.), *Loss and grief*. Basingstoke: Palgrave.
Midwinter, E. (1990). An ageing world: The equivocal response. *Ageing and Society, 10*(2), 221–228.
Moran, D. (2000). *Introduction to phenomenology*. London: Routledge.
Moss, B. (2007). *Values*. Lyme Regis: Russell House Publishing.
Moustakas, C. (1994). *Phenomenological research methods*. London: Sage.
Mullaly, R. (2002). *Challenging oppression: A critical social work approach*. Ontario: Oxford University Press.
Nelson, T. D. (Ed.), (2002). *Ageism: Stereotyping and prejudice against older persons*. London: Massachusetts Institute of Technology.
Nespor, J., & Groenke, S. L. (2009). Ethics, problem framing, and training in qualitative inquiry. *Qualitative Inquiry, 15*(6), 996–1012.
Nightingale, D., & Cromby, J. (Eds.), (1999). *Social constructionist psychology*. Buckingham: Open University Press.
Padgett, D. K. (1998). *Qualitative methods in social work research: Challenges and rewards*. London: Sage.
Peace, S. (2002). The role of older people in social research. In A. Jamieson & C. R. Victor (Eds.), *Researching ageing and later life*. Buckingham: Open University Press.
Phillipson, C. (1982). *Capitalism and the construction of old age*. London: Macmillan.
Phillipson, C., Bernard, M., & Strang, P. (1986). *Dependency and interdependency in old-age: Theoretical perspectives and policy alternatives*. London: Croom Helm.
Potter, S. (2002). *Doing postgraduate research*. London: Sage.
Randall, W. L., & McKim, E. (2008). *Reading our Lives: The poetics of growing old*. London: Oxford University Press.
Reid, K., Flowers, P., & Larkin, M. (2005). Exploring lived experience. *The Psychologist, 18*(1), 20–23.

Renzenbrink, I. (2004). Home is where the heart is. *Illness, Crisis and Loss, 12*(1), 63–74.

Riach, K. (2009). Exploring participant-centred reflexivity in the research interview. *Sociology, 32*(2), 356–370.

Ritchie, J., & Lewis, J. (2003). *Qualitative research practice: A guide for social sciences students and researchers.* London: Sage.

Roberts, G., Jones, E., & Ap Rhisiart, D. (2011). *Giving voice to older people: Dignity in Care Welsh Language Toolkit,* Cardiff: Welsh Government.

Rose, H. & Bruce, E. (1995). Mutual care but differential esteem. In S. Arber and J. Ginn (Eds.), *Connecting gender and ageing.* Milton: Open University Press.

Ross, F., Donovan, S., Brearley, S., Victor, C., Cottee, M., Crowther, P., et al. (2005). Involving older people in research; Methodological issues. *Health and Social Care in the Community, 13*(3), 268–275.

Roulston, K. (2010). *Reflective interviewing: A guide to theory and practice.* London: Sage.

Rubenstein, D. (2001). *Culture, structure and agency; towards a truly multidimensional society.* London: Sage.

Ryen, A. (2008). Trust in cross-cultural research. *Qualitative Social Work, 7*(4), 448–465.

Shaw, I., & Gould, N. (2001). *Qualitative research in social work.* London: Sage.

Smith, J. A. (2003) (Ed.), *Qualitative psychology: A practical guide to research methods.* London: Sage.

Smith, J. A. & Osborn, M. (2003) Interpretive phenomenological analysis. In J. A. Smith (Ed.), *Qualitative psychology: A practical guide to methods.* London: Sage.

Sunderland, J. (2004). *Gendered discourses.* Basingstoke: Macmillan.

Tehan, M. (2007). The compassionate workplace: Leading with the heart. *Illness, Crisis and Loss, 15*(3), 205–218.

Thompson, N. (Ed.), (2002). *Loss and grief.* Basingstoke: Palgrave Macmillan.

Thompson, N. (2009). *People skills* (3rd ed.). Basingstoke: Palgrave Macmillan.

Thompson, N. (2011). *Promoting equality: Working with diversity and difference* (3rd ed.). Basingstoke: Palgrave Macmillan.

Thyer, B. (Ed.). (2001). *The handbook of social work research methods.* London: Sage.

Ungerson, C. (2000). Cash in care. In M. Harrington (Ed.), *Care work. Gender, labor and the welfare state* (pp. 68–88). London: Routledge.

Varma, P. K. (2005). *Being Indian: The truth about why the twenty-first century Will be India's.* London: Penguin.

Warnes, A. M. (Ed.), (1989). *Human ageing and later life: Multidisciplinary perspectives.* London: Edward Arnold.

Webb, S. A. (2008). Modelling service user participation in social care. *Journal of Social Work, 8,* 269–290.

Whitbourne, S. K., & Sneed, J. R. (2002). The paradox of well-being, identity processes and stereotype threat: Ageism and its potential relationship to the self in later life. In T. Nelson (Ed.), *Ageism: Stereotypes and prejudice against older persons* (pp. 247–273). Cambridge: MIT Press.

Wilson, G. (2000). *Understanding old age: Critical and global perspectives.* London: Sage.

Chapter 5
Findings

Introduction

In line with my argument that research into reciprocity in eldercare has tended to neglect the broader sociological contexts within which it is experienced by older people dependent on formal care, and with reference to the research questions outlined in Chap. 4, these findings relate to the range of ways in which reciprocity can be seen to be impeded or promoted in such situations, and the levels at which such processes operate. As discussed in Chap. 1, these empirical data are not intended as the only means of addressing the research questions I set myself. Rather they, along with the literature reviewed, help to build up a picture of:

1. what it means to have become significantly dependent on others in old age;
2. the impact that such changes might have on one's identity as someone who may be needy in some respects, but can still contribute in others; and
3. the factors that have potential significance for the impeding or promoting of reciprocity in those circumstances.

I have used as much of the participants' direct narrative (or that reflected by the translator) as limitations on space have allowed. In doing so, I recognise that an element of my interpretation of their responses has been unavoidable. As is clear from my discussions relating to methodology in Chap. 4, I make no apology for that, as interpretation and meaning making are at the heart of the phenomenological paradigm, and the co-production of knowledge has been my aim. I consider the process to be, as described by Pascal (2006), one of a 're-authoring' of the participants' experiences, as my part in the production of the data cannot be divorced from that of the individuals I interviewed.

I begin this chapter by reminding the reader of the context of the research—very briefly revisiting the methodological paradigm and research methods that underpin the study—before moving on to introduce the 14 participants who shared their stories with me. I then report that, from a process of latent thematic analysis, there emerged 19 themes which I present below as they relate to: (a) social space; (b) social time; (c) meaning making at the level of the personal/spiritual and (d) meaning making at the level of shared meanings and institutionalised patterns of power. The design of the interview schedule has allowed for these findings to reflect something of:

1. the participants' perceptions of themselves as 'givers' prior to, and since becoming dependent on, formal care provision and
2. their understanding of whether others perceive them in the same light.

Before moving to the findings, I briefly revisit the focus and rationale of the study to set them in context.

The Study

As described in more detail in Chaps. 2 and 4, this study has both phenomenological and sociological underpinnings. The choice of a phenomenological focus was rooted in a desire to understand how older people attach meaning to their experiences and, given that the paradigm seeks to legitimise multiple perspectives (Becker 1992), for its potential to give voice to their often overlooked or trivialised viewpoints. Indeed, the themes I highlight below did emerge from that process of meaning making. Furthermore, it has a sociological dimension in line with my aim to theorise reciprocity in eldercare with recognition of the cultural and structural contexts in which meaning making takes place, and with reference to the dialectical processes operating there.

In order to generate data which might cast light on these processes, the same questions were asked of two groups of older people who had become dependent on formal care provision in the differing cultural contexts of Chennai, India and an area of the UK that straddled the North West England/North East Wales border. Seven individuals were interviewed in each location and I introduce them below.

The Participants

I am very grateful to the 16 people who agreed to participate in this study—two in the pilot study and the remaining 14 in the main study. My intention has been that they would be co-producers of new knowledge that would have the potential to inform debates that could contribute to beneficial outcomes for their peers, and for future cohorts of older people in circumstances similar to their own. From feedback I received after each person had read the explanatory letter accompanying the consent form (see Appendix 2), and from spontaneous comments from some of the participants in both countries, I am confident that most of them conceptualised the sharing of their stories and perceptions as acts of reciprocity—of giving back as well as being given to.

I present these brief pen portraits partly to locate the participants in terms of age, gender and health status, but also to acknowledge them as individuals with unique histories, aspirations and personhood. None of them voluntarily located themselves in terms of class origin, and I chose not to try to locate them as such, as the seemingly fluid articulation between class and caste in Indian society may have made any comparison problematic. As promised to them, their real names and places of abode have been anonymised.

The following women (all Hindus) lived in what they termed the 'destitute' section of Care Home A (situated several miles out of the centre of Chennai)—their care being provided by a small number of care staff, and paid for by charitable donation:

Lakshmi
Aged 109 years, Lakshmi, a widow with no children, had been a resident for approximately 5 years. She had been living alone with minimal assistance for many years prior to this admission, only moving into residential care after developing balance problems and breaking her leg. She had lived in several states in India and been a high school teacher and political activist earlier in her life.

Vadivu
Aged 78, and a widow with no children, Vadivu had lived in residential care for 6 years. At the time I visited, her broken leg was in the process of healing but she had a long history of arthritis, and balance and mobility problems meant that she was prone to falling if she moved about unaided. In her earlier life she had worked in the field of agriculture, as an educator rather than a labourer.

Subbulakshmi
Aged 73, and widowed with no children, Subbulakshmi had lived in the care home for five years. In her earlier life she had worked as a manual agricultural labourer (she had prided herself on her physical strength then) and had also worked as a cook and childcarer for a family who owned an agricultural company. At the time of interview she had developed generalised weakness, breathing difficulties and a debilitating loss of appetite.

In a separate section of the same care home, described by her as the 'hospital wing', and for which payment was required, lived:

Kannamma
Aged 70 and widowed, Kannamma had been a resident for 7 months at the time of the interview. Arrangements had been made by her son (living abroad) for her to move to the care home after she developed mobility problems. She appears to have lived most of her adult life in relatively affluent circumstances, predominantly as a homemaker.

And in a different care home (B), also situated several miles out of the city centre, lived.

Kumari
Aged 75 and widowed with children, Kumari had lived in the home for 2 years following heart surgery. As well as raising her children she had worked in the medical profession.

The two male participants from India (also Hindu) both lived in Care Home B.

Venkatraman
Aged 80 and a widower with no children, Venkatraman had lived in residential care for two and a half years. While in reasonable health, his ability to look after himself had been compromised by problems related to his blood pressure. He had

worked in marketing and management before retiring, and had been a trustee of his neighbourhood temple, a prestigious role in the community. He spent most of his days reading religious literature.

Muthu

Aged 78, widowed and with sons living elsewhere, Muthu had been a resident for 7 years. An asthma sufferer, he had spent most of his life doing secretarial-type work when his health allowed. His admission appears to have been based somewhat on choice in response to a dilemma he faced—he did not want to live alone, but felt he would be a burden to other family members if he became increasingly reliant on them as he aged. He used the term 'detachment from family' as justification for his admission.

While I am aware that all of the Indian participants were of the Hindu faith, this information was given to me by care providers, rather than by the participants themselves. I did not ask the UK participants about any religious affiliation and only one disclosed this dimension during the interviews. I now move on to present the UK participants, beginning with those who continued to live in their own homes.

Megan

Aged 88 and living alone, Megan was prone to falls because of a long-standing muscle-related condition. She had been offered care support on a number of occasions but had declined, preferring to manage the risks involved in order to lead the independent life that she craved and feared might be lost if she were to accept help. As well as raising her family, she had worked outside the home for much of her life and been a keen sportswoman. Widowed with sons living elsewhere in the country, she was keen to continue to be as involved in her local community as possible.

Harold

Aged 84, Harold was a widower sharing a house with a daughter who did not act as his carer. Because of a progressive neurological condition, he needed help to wash, dress and cook and was helped in these respects by two carers who visited daily. He had been retired for many years from his job as a teacher.

Ken

Aged 82, Ken was very debilitated by a number of strokes and kidney failure. Living at home with his wife (who had her own health problems) Ken had been supported by daily visits from a care team to help him to wash, dress and travel to hospital and daycare centres. He had enjoyed his career as an engineer and had also taken pride in the care and repair of his home before becoming ill.

The four remaining UK participants all lived in the same residential care home (C).

Doris

Aged 90 and widowed with daughters, Doris had moved into the home on a private basis after a collapse due to a weak heart had given her family cause for concern, and had undermined her confidence in living alone. For the majority of her life Doris had been a homemaker and had taken pride in that.

Maureen

Aged 70, single and with a long history of mental health problems, Maureen had become resident in the care home after the death of her mother. Without that mutually supportive relationship she had become depressed and lacked confidence to live alone. In her earlier life she had worked outside of the home when her health allowed, but this had been intermittent employment and mostly earlier in her life.

Edna

Aged 79, Edna had been living there for 3 years, after increasing giddiness caused by heart problems had made living alone at home very difficult for her. Now widowed, she recalled the joy of raising a son, while also enjoying her job in the library service.

Eddie

Aged 77, Eddie had been resident for about a year, after becoming unable to look after himself following complications of diabetes, and heart problems. He was divorced with grown up children and had worked in psychiatric institutions prior to retirement.

I thank them all for their vitally important part in this research.

The Themes

In earlier chapters, the significance of social time, social space and meaning making for an understanding of interconnectedness and intersubjectivity in social life has been established. Using these concepts as a pre-defined framework, I analysed the interview data to assess the extent to which these concepts featured in the narratives of those who were interviewed.

The transcripts of all interviews were scrutinised and, where a concept relating to any of the key phenomenological categories of social space, social time, or meaning making was identified, it was coded as such and recorded under the appropriate heading. When all of the transcripts had been coded it became possible to identify sub-themes as follows:

1. *Social space (relationships; operating in a social world; connectedness)*

1.1 Place-in-the-world: self-perception as a valued member of society
1.2 Perception by others as a valued member of society
1.3 Perception of opportunities to reciprocate
1.4 Expression of the desire to reciprocate
1.5 Relationship between geographical and social space—barriers and enabling spaces
1.6 Understanding and expectation of care relationships.

2. *Social time*

2.1 The future dimension/personal growth
2.2 Self-perception as changing in terms of being a valued citizen
2.3 Expectations of old age in India and the UK—continuity and change

2.4 Intergenerationality—reciprocity between generations
2.5 Developments in communication technology—opportunity and barriers.

3. *Meaning making at the level of the personal/spiritual*

3.1 Have I got anything to give?—recognising reciprocity where it exists
3.2 A sense of being valued rather than being a burden
3.3 Dependency as existential crisis—am I the same person or someone different now that I need some help?
3.4 Giving back—duty or pleasure?

4. *Meaning making at the level of discourse and institutionalised patterns of power*

4.1 Health discourses—old age as illness
4.2 Social construction of old age
4.3 Discourses around dependency: care model and overprotection
4.4 Welfare policy and practices.

These themes obviously held significance for the participants because they were derived from their processes of meaning making but, in order to get some sense of how significant these themes appeared to be for them, a count was made of the number of times they had featured in the transcripts. These levels of significance are reflected in the findings as articulated by the elders themselves and presented below.

The Participants' Perspectives

As can be seen from the interview schedule, I had incorporated a temporal element to the guide questions because, in addition to whether their own perceptions matched those of other people's in the care relationship, I was interested to know whether they had felt differently about themselves as 'givers' since having the label 'dependent' attributed to them. The first part of the interview schedule, then, sought to elicit the participants' experience of being a 'useful' or valued person throughout their lives prior to becoming dependent on formal care delivery—that is, on how they had constructed their identity as a multifaceted person. As the term 'reciprocity' does not tend to be used very much in everyday speech, the questions and prompts were framed around concepts like pride and self-esteem, such as 'what made you proud?' or 'what did you consider yourself to have been good at?' To some extent then the responses related to what people *did*, rather than how people *gave* but, given that I have made it clear that I use the term reciprocity to imply a sense of 'usefulness', then I would argue that the 'giving back' element is often incorporated in the 'doing' aspect, and so it would be unnecessarily arbitrary to distinguish between the two in people's responses.

In all cases, the participants needed several prompts before they were able to begin to talk about themselves as having had something to offer others—to think of themselves as 'givers'. For example, prompts included 'was there anything you used to do which made you feel good about yourself?' and 'was there anything that made you feel proud?' However, once prompted, all were eventually able to describe experiences that had contributed to their self-esteem and self-worth, even though they had not previously conceptualised them as anything more than what was expected of them in their respective social roles.

The questions then moved to focus on whether, in their current situation, they still thought of themselves as someone able to give as well as receive—that is, to *maintain* their identity as a multifaceted person. Perceptions at this stage were gauged by responses made in the course of conversations prompted by the questions in the semi-structured interview, but also from their responses to the short questionnaire I administered as part of that schedule at the end of each interview.

This far, the focus had been on the participants' *self*-awareness, but I had been keen to hear what they had to say about how they felt they were perceived by others. An element of the schedule therefore focused on whether they thought that those involved in supporting them currently saw them as someone able to give as well as receive—that is, as a multifaceted individual, or as the one-dimensional person enshrined in the stereotype of old age as stagnation or decline.

The excerpts I present below provide a degree of evidence to suggest that older people dependent on formal care are often capable of reciprocating, and want to do so, but feel that this is often not appreciated by others who are in a position to help them fulfil their ambition and potential. It is to the first set of responses, those relating to where and how people live their lives in relationships with others—their 'being-with-others' that I now turn.

Social Space

Place-in-the-World: Self-Perception as a Valued Member of Society

General. This appeared to constitute a particularly significant theme, characterised by concerns about having had a valued role earlier in life (teacher, engineer, homemaker, for example) but having lost that status since becoming classified as ill or frail. Also incorporated in this category are references to pride in either their former, or current, capacity to act as givers.

Specific. For several in the study, the passing on of wisdom in the form of knowledge or technical expertise had been the source of a sense of pride in being able to give to others. For Lakshmi and Harold this was expressed through their roles as qualified teachers. For example, during the interview with Lakshmi, when I asked whether she had worked with children she replied, while pointing to her head,

> Oh yes, lots of knowledge' good knowledge ... I was at projects and very good teacher, teacher of Hindi.

When she had reverted to speaking in Tamil, the translator explained that Lakshmi was reporting that:

> The teachers and professors paid respect to her and made her feel proud.

Harold, also a teacher, was more self-deprecating about what he had been able to offer, although he did make the point that he had returned to North Wales after teaching in England so that he would be able to teach children to speak Welsh, his first language and so an area of expertise.

A similar source of pride was also reflected in less formal educational scenarios. Kannamma, for example, described how she had interpreted her role as a mother as helping her children to have the best opportunities in life, but also took pride in offering tuition in maths and Hindi to other people's children. For Doris, a sense of giving was expressed in remembering having her daughter learning alongside her when she was sewing all day at her home, and in making dresses to raise money at church fetes. Kannamma also expressed a sense of pride in being able to pass on techniques relating to kolam, the traditional floor decoration made with rice flour and associated with the festival of Pongal.

Both Subbulakshmi and Vadivu, referred to having been able to pass on knowledge and techniques relating to farming. Referring to working with her father after being widowed, Vadivu commented:

> Yes, agriculture. I went to look after the ladies department, he would look after the men's department. We were very happy … transplantations, seed looking-after and harvesty … what we knew we taught to farmers.

Supporting the learning of others was also referred to by Venkatraman, in his role as an office manager:

> Because the world is so geared to the highly educated I try to make people as educated as me.

His wry sense of humour led me to wonder whether this response was somewhat tongue-in-cheek, although his projects and priorities later in life suggest that education was something he valued highly, and so may have been serious about getting a sense of pride from supporting the learning of his staff.

For Edna, literary knowledge was declared as something she felt good about being able to pass on to others, as too was expertise in dietary and health matters for Kumari. On a more practical level, Ken expressed great pride in his skills as an engineer, and what he could offer to his employers in particular. Having described his work designing air compressors and valves, he mentioned with great pride that:

> I was that skilled—well, when I was a fitter on the bench, I was only 20 and got paged. I had to go to the managing director … to America for two months studying production and from then on, I never looked back.

He went on to describe a flourishing career in which he moved from being a development manager to a service manager, and a joy in and devotion to his work such that he took only 1 week off sick in the 51 years he worked for that company. After retirement, but before becoming ill, Ken's expertise remained a source of pride for him:

> When I first left work, when I was 65, because I was that good at me [sic] job I ... we used to have ... because we made valves and things, we used to have them sectioned so that they could display them. And they used to pay a lot of money for them but they said to me—"well, look now you're retired, would you like to do this?". I used to do all the section models here at home. I had a little workshop upstairs and I used to section the models and get paid for it, you know. And then I started DIY.... I'm good at woodwork, electrics, mechanical, plumbing ...

Several of the participants referred to the sense of purpose they experienced through having a designated role from which they derived status, and performing it to the best of their abilities. For the men in both samples this was reflected in being able to support their families as breadwinners, although both Venkatraman and Harold also referred to contributing in the domestic sphere, as both of their wives had also been working outside of the home at the time. Doris and Megan both referred to being committed to their roles as homemakers, and being proud of what they had been able to give to their children in terms of time and support, as too did Kannamma, who spoke of only wanting praise for her investment in her son's interests;

> My son told me you look after me well. For me there's a lot of work you are doing, so take rest ... [but] my father told me my son came first—this is blessing—so I only want praise from my son.

Several of the women also spoke of the sense in which they felt proud of being able to contribute something to the public, as well as the private, sphere. Megan spoke of seeing herself as the family mediator in disputes between her husband and their four sons, but also as playing a competent role in the workplace. She remembered being called on by a former boss to do some work for him, which invoked a memory of her then colleagues' perception of her as competent:

> The girls used to say when I used to turn up, you know, thank God you're here Megan. We had a terrible woman here last week. We couldn't do our own work because she was asking us all the time and they used to say, don't worry, Megan's here next week.

Maureen's perceived sense of loss at having had no clearly defined status either in the public or private sphere, was evident throughout the time I spent talking with her. She described having had no sense of purpose in her life after leaving the job she had held at a residential school many years before, then having been out of work since, and given that her role as her mother's carer had ended.

> It was good to do a job and keep involved with people and getting on with things. And when I left there, I just went downhill really. I'm not very good and feel down and don't feel worthwhile as a person really ... I'd like to have been a mother and brought up children ...it never happened. I had good friends but I sent them off and things like that. I didn't go to their dates or anything like that so I feel useless, really.

Perception by Others (now) as a Valued Member of Society

General. Given that several references were made to not being listened to in general, or asked by care providers about their strengths, it seems that the participants had significant concerns about the extent to which they were valued by others in their current circumstances.

While all of the participants had something to say in respect of their own perceptions of their 'usefulness' prior to their current situations of significant dependency, the response to this line of questioning elicited fewer and less developed responses than the previous one had. This is perhaps understandable given the fairly short timescale of the interviews, and that I was asking for a degree of introspection and analysis that ideally needed more time to be developed. In retrospect, had the opportunity been available, I might have considered conducting a second interview during which I could have used more prompts to help the participants to reflect more deeply on their perceptions of care delivery. Nevertheless, the interviews produced some enlightening insights.

Specific. For example, several people said that when they *had* been perceived as having valued roles, it had been by fellow residents, rather than by eldercare practitioners, or care facilitators. This appeared to be the case with Vadivu's prayer leading, and the informal approaches made by the young women students to Lakshmi and Kannamma, for help with their studies.

A lack of recognition at an organisational level of older service users' desire and competence to reciprocate, or the facilitating of opportunities to do so, was also evident in Harold's comment that he might have been able to help fellow sufferers of his medical condition, by sharing his knowledge and experiences. He could potentially have done so in a semi-formal capacity, such as the 'expert patient' role described by Thomas and Moncrieff, as cited in Bytheway et al. (2002), but in Harold's case, this happened only coincidentally—if, as he said, '*it came up in conversation*'. Similarly, Muthu described feeling good about being able to help his fellow residents in, as he put it, '*sundry ways*', but his willingness to be involved in charitable work at a more consistent and organised level was not facilitated.

When I used the checklist of potential areas or scenarios in which they might have been able to contribute, four of the participants made reference to the fact that many of these were organised or carried out by staff anyway, so that there was no call for the participants to offer their services or expertise. For example, at the time of interview, Venkatraman, having previously managed a music society, now just attended recitals organised by staff. And Maureen, Muthu and Doris all commented on how *residents'* meetings at their residential home were arranged and led by *staff*. Having been told that one of the managers organised residents' meetings, I asked whether the residents played any part in, for example, agenda setting:

> Researcher: In terms of planning the day? ... what you're going to talk about?
> Maureen: That's all done by the office ... we don't plan what we're going to say. X just asks questions and we make suggestions and if it's a good suggestion she acts on it.

Doris described what took place in residents' meetings as '*half-hearted*' and the only instance she could recall of being offered an opportunity to have her perspective heard was when asked by one of the managers about her opinion on menu choices already made by a third party. I, too, would regard this as a tokenistic response to user involvement when analysed in relation to the participation continuum referred to by Hickey and Kipping (1998).

There was some comment from Kumari and Eddie, however, which suggested that they felt their competence *had* been recognised to some extent. In Kumari's case it was with reference to the funeral arrangements she was entrusted with, and Eddie referred to a recognition of his skills in 'people work':

> Well, the lady here [referring to the manager]—I think she's keen on getting me back into the swim, you know. We have, how can I put it? ... a clients' committee, I call it ..." Would I be interested in, for want of a better term, using the skills I've acquired?". I can move around with most people because of the profession I was in.

However, while he felt he was perceived as competent to play a part in that particular project, he commented that some of his carers were dismissive of what he perceived to be 'useful' feedback about policy and practice at the home in general—a judgement that affected his sense of self-esteem. For example, in one situation where he had commented on changes in guidelines relating to the administering of medication, he reported feeling affronted by the implication that he would have nothing of consequence to say, having now retired from the nursing profession:

> I think "what the hell do you think I am? Some kind of fruit and nut case?" I feel as though, at times, it's been on the tip of my tongue to say "who the hell d'you think you're talking to?"

Subbulakshmi also referred to attempts to limit her attempts to contribute in her own way to her immediate environment (by cleaning the communal entranceway), supposedly in the interests of protecting her from perceived physical harm. She did not refer specifically to how this may have had an impact on her self-image as someone with the capacity to reciprocate, but the fact that she continued to clean the grounds and entrance porch surreptitiously, and against advice, suggests that this may indeed have been the case. Although *she* appeared to have been perceiving herself as competent to reciprocate, it seems that her spiritual affirmation as a useful person had been sublimated to risk-management concerns on the part of her care providers, who appeared to see her differently—as in need of protection, rather than in need of support to continue doing what made her feel valued.

Conspicuous by their absence, in all but a few cases, were responses which suggested that care practitioners or facilitators had asked the participants about future aspirations to continue being the person they had always been, or wanted to become. Maureen remembered being asked at some point what she wanted to do with the rest of her life, but nothing further appeared to have developed from that initial conversation. Harold's experience seemed similarly unfocused and lacking in commitment to action. For example, during a needs assessment process, he remembered talking about whether he could be helped to develop any outside interests he may have had, although he told me that this aspect of need had been raised by him, rather than the person leading the assessment process, and that no help had been forthcoming anyway. By far the majority of participants either had no memory of aspirations being explored at all, or were absolutely sure that they had not been on the agenda.

I had asked the participants whether they thought that their gender had played a part in the extent to which they were perceived by their carers as able to 'give back' in their situation of being dependent on formal care. Most struggled to respond to this, which was unsurprising but disappointing. As I did not have the luxury of time to probe more deeply, I left this line of enquiry relatively undeveloped and may take it up again independently of this project, at some future point. However, Lakshmi's response spoke specifically about the patriarchal nature of Indian society, and the relationship between gender and age as social divisions:

The translator fed back that:

> She is stating that, even though old age comes, the people here respect old age and they take care of old people;

but then went on to say

> so she says there will be a very big difference if she would have been male because she would have got more respect than what she is getting now—it is always male dominated.

While Lakshmi's was the only explicit reference to gender differences in eldercare, the gendered nature of old age in both countries was implicit in some of the responses to other questions.

Perception of Opportunities to Reciprocate

General. There were enough references to either perceived opportunities to reciprocate (such as mutual collaboration between care home residents, or the extent to which expertise was drawn on) to suggest that this was a particularly significant theme.

Specific. It became clear during the interview process that perception of opportunity to give back was, in part, dependent on how the participants defined reciprocity. In terms of my own reflexivity, I realised that, when I explained reciprocity to the participants as 'helping others', I had in mind reciprocity in whatever form it took—incorporating, for example, practical or moral support. However, I had not conceptualised it as incorporating financial help. It became clear from several of the responses that it did have that connotation in Indian society—or Chennai at least. For example, in saying that she was not keen to 'help' others, Lakshmi referred to *'not indulging her sympathy'* for fear of showing favouritism among the residents, which had the potential to result in what she referred to as the *'fighting and falling out'* that could have tarnished her good name. The translator confirmed that Lakshmi's reference to 'indulgence' did refer to the giving of financial support. Kannamma's responses also suggest that she interpreted 'helping others' primarily in a monetary sense, as she appeared to perceive giving as involving material goods:

> Since I came here I brought things—garnets, turquoise, diamonds and gave to everybody—tops, skirts ... Yes, my son has spent 100 dollars ... they will know my son only brought. *Me*, I will give.

While she referred enthusiastically to her being able to help others in that sense, any reference to other forms of helping were drawn from her life before becoming dependent, or significantly underplayed as not being 'expected' because of her status as a 'paying guest'—a point to which I return to Chap. 6.

While Kannamma appeared to have perceived helping others as being in her control, this was not the case for everyone. Muthu, for example, relied on his aspirations being facilitated by others. He perceived reciprocity as being able to support those less fortunate than himself—in effect, a moral obligation. He had suggested that involvement in what he referred to as 'social work' (more readily understood in the UK as voluntary or charitable work) had always been important to him, and that he would have liked the opportunity to continue to be as involved as his current health and circumstances would allow;

> At one time I could easily do it if physically and mentally healthy, but nobody has asked me to do social work.

I asked whether he would like to do it now. He replied;

> Yes, but nobody here asks. There is very small opportunity here, because everybody who is here is not asking. A little bit I do, but if anyone was asking me to do social work I could do more.

Ken and Eddie also appeared to feel that opportunities for what *they* defined as reciprocity were lacking because they were not recognised as such by others, and frustration at not being able to contribute in the ways they would have wished was a feature of several of their responses. Conversely, where *I* identified acts of reciprocity (such as in Harold's 'gift' of friendship to one of his support team), they had not always been recognised or defined as such by the participants.

Megan, appeared to actively look for opportunities to reciprocate—as she said:

> It's in you—you know?

She described taking on advocacy roles, both informally (when she came across what she perceived as injustices) and in response to a formal invitation to act as an ambassador for the charity Age UK. However, some of the participants, while they may have responded reactively to requests for help, did not seem particularly proactive in seeking out opportunities to reciprocate. Doris, for example, appeared less willing, or able, than others to express ways in which she felt she was able to give back to other people, or her community. Many of her responses to questions about her current situation referred to being bored, but she did not appear committed to doing much about that. Eventually, with prompting, she expressed a perception of herself as a good but reluctant listener. When I asked whether she thought she was still able to give help as well as be helped, as she had reportedly done very competently in her earlier life, she commented on her interactions with fellow residents;

> Well I do when I'm talking to them really, but some of them … well, they all seem to take me that I'm there for them to unfold their troubles.

When asked how being able to give that time to others who appeared to need her support felt, she replied;

> Well, alright, except that they do rather put it on, you know, put on me and you feel that you're getting fed up with it.

Although describing herself as being bored, Doris expressed a reluctance to become involved in opportunities that others saw as having the potential to make good use of skills of which she had formerly been proud:

> I couldn't believe my daughter said at the weekend, you ought to be doing the minutes and typing [at the residents' meetings]. I said don't be so silly, I can't do those things now … but I must admit I loved typing. Don't know what I'd be like if I got on one now. I don't know, it's a long time.

Similarly, in response to the checklist questions relating to an interest in opportunities to give back in some way, she expressed either no interest in them (other than the opportunity to '*teach modern children good manners*') or told me that the opportunities for using her domestic expertise in her current situation were not forthcoming because others were employed to manage such matters.

Maureen's response was interesting in that she appeared to link opportunities for giving, or giving back, almost exclusively with a work status, so that her sense of self-esteem was necessarily linked to having a job and a self-sufficient life. Though she had not had paid employment for many years, it seems that she had derived most of her sense of self-esteem from having a job and living independently—something that had defined her as 'myself', but which she no longer had:

> Sometimes I feel as though I'd like a job but I'm too old and, other times, I think I'd like a little place of my own, back to myself again.

I had wondered whether she might therefore have sought to create a work-type role for herself at the care home—perhaps accepting, or even seeking out, roles and responsibilities which may have helped her to maintain the old sense of 'myself as valued', or to create a new self-image of which she could be proud. However, when we talked about what she might like to be doing with her life, she expressed only a wish to be left alone to 'do her own thing' (which appeared to amount to sitting alone in the porch much of the time) and offered little that spoke of a perception of herself as someone who had something to give;

> I'd rather be by myself and just toddle along… I feel happy that way.

When I focused more particularly on opportunities to reciprocate, by using the checklist questions as prompts, there was still little evidence of a desire to be perceived as someone who wanted to be involved in the residential community that had become her home. She claimed that no one ever asked her to help with anything and, when I asked how that made her feel, she replied;

> I don't mind. I just go by what people say … as I said, I'm just a resident here so I just go by the book really.

When prompted further, it appeared that she had once been asked to help in the office but had declined. Of the little bit of washing up she occasionally took on as a task, she said that it made her feel good;

> I've done something, got off my backside and done something instead of sitting all day.

It appeared from this, and other responses, that it was having something to do, rather than having something to contribute, that made her feel good. For example, although Maureen said that she occasionally talked with other residents, her responses suggested that this was perhaps more in her own interests than theirs:

> You just talk and say things and commiserate … it's a relief really … I like talking and most people just and go to sleep here.

Expression of the Desire to Reciprocate

General. This was a broad and, it seems, particularly significant theme. Examples of responses of this nature include the extent to which people felt qualified or confident enough to put themselves forward as someone who has something to give, and their perception of whether this was appropriate behaviour in relation to their age or culture. It incorporates too a perceived link between reciprocity and well-being, and also aspiration to continue living a life underpinned by cherished values.

Prior to the interviews I had compiled a checklist of prompts (see Table 5.1) as an aid to generating discussion about the ways in which the participants might have aspired to be of use to others, and whether opportunities for such had been facilitated by care providers. I had constructed the list according to what I thought might have been implicated, in the hope that this might have prompted the participants to add their own examples, thereby reflecting their lived experience more directly. Unfortunately, while everyone was willing to comment on those examples I had chosen, no one added anything of unique relevance.

Nevertheless, the results were interesting in that they highlighted that, over the ten areas commented on:

1. as far as the Indian elders were concerned, the relationship between what the participants would have welcomed being able to help with, and the extent to which they had been able to, was fairly evenly matched in most cases. However, this does not necessarily suggest that they had been *en*abled to do so by eldercare practitioners, and may have reflected the cultural expectations of self-responsibility on their part; but that
2. reflected in the responses of the UK-based elders was a more marked disparity between what would have been welcomed, and what the participants had been able, or enabled, to do. In all but one category (the providing of emotional support, which, I would suggest, is likely to have been self-initiated anyway) the expression of desire to help others in the ways identified was higher than instances of it being enabled.

Specific. Some of the participants overtly expressed a link between reciprocity and well-being, suggesting that they sought out opportunities for that reason. In Muthu's and Edna's cases, for example, this was a positive link. Muthu's motivation to give to others through charitable works may have been underpinned by a sense of duty related to dharma—the set of normative rules that assigned him obligations relevant to his station in life (Jacobs 2010), and something that was instilled in him by his older brothers in line with the family's world-view. He intimated this in the following comment:

> God has created me in that light, that temperament to help others.

And, when asked whether being able to help others made him feel good, he replied:

Table 5.1 Results from interview schedule (see Appendix 1)

Potential opportunities for reciprocity	Indian elders		UK elders	
	Would be welcomed	Enabled	Would be welcomed	Enabled
Organising meetings, rotas, etc.	2	2	6	3
Involvement in training of staff	6	4	6	2
Teaching children and young people	6	3	5	2
Cooking/advising on recipes/diet	4	3	3	2
Mending/repairing and advising on such	3	1	5	1
Plantcare/gardening and advising on such	2	2	2	0
Emotional support/befriending, etc.	6	6	4	4
Home-making and advising on such	3	3	5	2
Advising on health and coping with illness	6	6	3	2
Keeping informed re politics, current affairs, etc.	5	3	2	1
Other	–	–	–	–

> Yes—so to go to paradise,

which he appeared to presume would be his reward for having lived a 'good' life, or succession of lives, in the service of others.

Edna, too, identified helping others as a 'feel-good' factor prompting a desire in her to engage in such activities:

> Oh, I'd break my heart, then [if no-one came to ask her advice]—it would break my heart then... well, you see, I'm that sort of person, although I say so myself. I'm very fond of people. I love helping people.

Similarly, where Ken found opportunities to 'give back', he found it life affirming to be perceived as someone who still had something to give. When prompted to think about ways in which he perceived himself as still being able to contribute, despite being compromised in some ways, he described passing on his expertise to others. This included his family:

> Well, I mean, I've taught me daughter to decorate and that, and they decorate for me ... and I've got grandchildren that are always in me garage with me, or the shed and they can all use the tools, you know. So I've been teaching them all me life.

But also friends at the club where had volunteered his skills before he had become ill:

> Oh yes, they will [come to him for advice]... the chap that does DIY there now—I've had quite long chats with him, you know, because most of the stuff that's been done, *I've* done, so he can follow on from what I've done ...

And when asked how being able to contribute in these ways made him feel, Ken replied;

> Smashing. Makes me feel really nice ... they see me as a person and not as disabled, you know.

Eddie's narrative, however, spoke more of his well-being being compromised by what he perceived to be a lack of respect for his expertise, and barriers to his ambition to have a valued role in the community. During a conversation about managing risk, he commented:

> I think the way some people attend to the caring of the elderly, myself included, they override the person. I know they've got government orders and all the rest of it, and I know where I'd like to put those at times!

For some, the desire to reciprocate may have been underpinned by what Grenier and Hanley (2007) refer to as 'acts of resistance'—a point to which I return in Chap. 6. Megan, for example, clearly did not want to be defined by her dependency and seemed to regard acts of reciprocity as a way to disassociate herself from what she perceived 'old age' as being. Similarly, Lakshmi also challenged ageist-influenced perceptions of what most people would expect of a 109-year-old woman by continuing to be a 'giver'—using her experience as a teacher to help the younger women whose education the owner of the residential home was sponsoring. As reported by the translator;

> She says that she has the education and knowledge she has—through that she can use anywhere, anytime.

Her responses to the checklist questions reveal that Lakshmi would have welcomed, given the chance, a broad range of opportunities to help, including:

- contributing in some way to the training of health and care workers;
- continuing to be involved in the teaching of children and young people; and
- using her experience and knowledge in the interests of the physical and emotional well-being of others in her community.

But while Lakshmi's desire to reciprocate in whatever way she could was shared by most of her fellow residents, there were some participants to whom being seen as a 'giver' as well as a 'taker' did not appear to be particularly important. Maureen, Doris and Vadivu had all been resistant to the proposition that they had the potential to reciprocate, but for different reasons, to which I return in Chap. 6.

Relationships Between Geographical and Social Space: Barriers and Enabling Spaces

General. While it did not feature particularly strongly in the narratives, it was nevertheless significant, in that there was a reflection in some of the responses of physical and ideological barriers to reciprocity, people's desire to be perceived as valued, or their projects to 'stay me'.

Specific. Venkatraman, for example, could not play a part in temple life, as he had done before becoming dependent, because he did not have the means to travel to the temple from the residential home. Similarly, Eddie was isolated from the community in which he felt he could play a 'useful' role, by not being able to travel there from the residential home without the assistance of staff members. Not only that, but also, as he perceived it, by the attitudes of those staff members. While accepting that the potential for him to fall over if unaccompanied was fairly high, he appeared to prioritise his self-esteem and spiritual well-being over the risk of physical injury. He had already said that he felt isolated from the community in which he could have been playing a useful role and added that:

> Well … I'm not a doddering old man just because I'm 77. I like life too much to give it up just yet;

which suggests that his definition of what constituted 'life' may have incorporated a sense of well-being derived from feeling valued, that was not so significant in the eldercare practitioners' perceptions of what should constitute 'life' in old age. As a result, then, the only environments in which he, Venkatraman and their fellow residents were 'allowed' to operate were those where a dependent status was the norm, and reciprocity not an expectation.

I had been interested to know whether the developments in communication technology (especially the Internet, Skype and networking sites) that appear to be realising their potential to redefine social space and the way relationships operate within it, (Hudson 2002), were being recognised and embraced by the participants. With regard to the Indian elders, none had been given access to any such technology, and nobody made any reference to how it might have been in their interests to have done so. Where reference was made to it by the UK elders, it was either as a communication medium with family (as in Megan's case), or in relation to the office administration of Residential Home C, in which some of the residents had been invited to play a minor part, but had declined. As with the Indian participants, none of the UK elders referred to its potential for opening up new avenues for reciprocity or personal growth.

Understanding and Expectations of Care Relationship

General. There was some evidence of an understanding of care relationships, and their significance for what people perceived as expected of them within those relationships, but not all of the participants had the same perception of the nature of the relationship between eldercare practitioners and themselves.

Specific. Maureen had spoken of not feeling the need to reciprocate because she perceived herself to be in residential care to be 'looked after' and needed only to 'play by the home's rules', as she saw them. That is, she appeared to be adopting the submissive role she felt she *should* be playing as a care home resident (one that did not incorporate giving back), and would continue to play until she was deemed fit to return to her own home and the 'myself' that she felt she had forfeited:

> I came here to be looked after, really… I'm tied to the psychiatrist—it's up to her what I do.

To some extent, Harold appeared to have had the same perception. When I asked him whether his care team had been told about, or sought to know about, those aspects of his life that were important to him he had replied:

I don't suppose they need to know. It's just somebody to be helped, that's all.

However, though at one point he had described himself as '*past his sell-by date*', he nevertheless tried to guard against his relationship with his care team from becoming too one-way, by trying to maintain some of his independence:

They can do sometimes [try to take over]—you know suffocate you, if you let things go that far,

but also by offering something as a counterbalance. This was evident in the way he described his relationship with his cleaner as friend to friend—one in which they felt comfortable to offer each other support through difficult times, such that he said of their relationship:

It's mutual isn't it ... she doesn't regard it as a chore coming here. And I look forward to her coming.

Vadivu, like Maureen, spoke of an acquiescence to 'being done to' by her carers—of not seeing it as her place to concern herself about the needs of others, or reciprocate for care received. But, in making the following comment, Vadivu may have perceived 'helping others' as necessarily meaning physical help:

Now I can't do anything—and someone should do for me ... No they never asked that [the positive things that she feels she could still do] because I am handicapped. They won't ask me to do any work. They will help and I will rest.

Others' perceptions, however, provided a challenge to the assumption that becoming dependent on formal care negates any competence that a person might retain, and wish to use for the benefit of others. For most of them, this challenge appeared to be associated with not wanting to be, or be seen as, a burden. Muthu expressed this overtly:

I'm not a selfish person... Helping tendency is not to be a burden. I want to be active.

But it seems also to have been implied in the responses of others, such as in Lakshmi's hiding that she had fallen and suffered pain from her knee injury (for fear of being denied her independence), and also in Subbulakshmi's continuing to contribute to the upkeep of the residential home in ways that had been characteristic of her contributions to households throughout her life, such that she continued to refer to them as 'her duties' even when no one had those expectations of her.

Reflected in the responses a number of times was the perception of the care relationship as teamwork. Eddie, for example, described a two-way process, of which he thought of himself a part:

To be honest, I like to think that, in a place like this, I contribute to the well-being and, funnily enough, they contribute to my well-being. I mean both staff and clients.

And, on more than one occasion, Kumari used the term 'mutual' in her responses—describing calling on experience or knowledge, and helping each other through difficult times, as acts of mutual giving. Her responses reflected a pride in being able to contribute to the well-being of fellow residents in a number of ways, including practical, spiritual and emotional dimensions, especially in times of health crises or death:

> In our home, many difficult and emotional events has happens [sic]—on that, I will phone to the doctors—making arrangement for the funeral van and the death certificate and inform the person's relatives ... will give physical and moral support at this critical time. This is one of my best services to the home.

And while not relating specifically to the relationship between carer and cared-for, underpinning Venkatraman's responses was a perception of giving back as being part of a mutual support network.

> Here, I am helping. Everyone is co-operating. So, if I want something they [points to fellow residents] will come and help me and if they want, then I will help them also. Yes, we will know. We are well-suited. Like family ... I feel good and proud of me.

Teamwork, however, did not feature in Kannamma's perception of the relationship between herself and her carers. Rather, she described it in terms of that between a provider and paying guest. For example, although proud of her handicraft skills, when asked whether she would consider reciprocating for care received by the giving to staff of handmade gifts (as Subbulakshmi had also done), she replied:

> Definitely not! And why? We are paying you know!

Becoming indignant, she continued in Tamil, which was reported as;

> She is saying that it is not their place. The staff will not come and ask her to make things because there is a distance. If the same thing would have been with the rescued elderly then they'd have gone and asked maybe to do things with them, but she's staying in the paid part so they do not come and ask.

I considered that her response may reflect consequences of the 'commodification' of care (Ungerson 2000) and return to the significance of this in later chapters.

Social Time

Future Dimension/Personal Growth

General. This was a very significant theme, as is apparent from references to a fairly imminent death and, in the case of some of the Indian participants, a deference to their god's will about that, and the changing role they should play as they age. It incorporates, too, a desire by some to continue to operate in the present and future as they had done in the past—a sense of life-long ambition and growth, but also a feeling that these aspirations for the future are not always recognised.

Specific. For the most part, though, the narratives focused more on the present and the past than on the future, except for a few references to remaining healthy enough to be able to continue to play 'useful' roles, or to continue on a path of personal development. In Kannamma's case, this referred to the hope that she would keep healthy enough to be able to reciprocate:

> While I live I want to do something—I want to help somebody. So I'll pray to God that hands and legs are fine.

Megan appeared to be equally concerned to continue living a 'useful life' but was more proactive in taking steps to minimise what she perceived as a potential physical and intellectual deterioration that might occur in her soon to be 10th decade of life. Describing being offered care support she reported:

> So, no, they offered me everything you know, and they said 'is there anything else we can possibly do? 'I said I can't think of anything because I've got to keep moving.

Her insistence on getting out and about despite her frailty, and her daily programme of crosswords and solitaire, suggest that she may have seen keeping active as a way to hold incapacity at bay.

But while the future may have been on the participants' minds, what featured significantly across both sets of responses was a lack of attention on the part of those providing support to them, to the participants' aspirations for the future. As mentioned in point 1.2, only Maureen and Harold had any recollection of having discussed with carers or care facilitators whether they had plans for the rest of their lives. And, even in those cases, fulfilling those plans had still been left in the hands of the participants themselves to organise, suggesting that the spiritual need for maintenance of a valued identity and personal growth had, for most of the participants, been unrecognised, or considered outside of the remit of assessment of, and response to, need. It seemed, for example, that Harold had always benefited greatly from attending a variety of evening classes, and would like to have continued to do so in order to reciprocate by contributing to discussions, and to continue on a path of learning and personal growth. Lack of transport had been the only barrier to fulfilling that wish, yet it was not forthcoming from care providers or facilitators:

> I used to go to night classes ... I organised that myself. Only I need transport to get there.

And Eddie's conviction that he, too, had the potential to carve out a new role for himself—building on past experiences to create a spiritually fulfilling future as a counsellor, advocate or friend to aged ex-service personnel—also appeared to go unrecognised as a need. Furthermore, it was a need that could easily have been facilitated, especially with the advances in communication technology that could have allowed him to operate this service from within the residential home.

> I was a national delegate with the British Legion ... and I really enjoyed that side of my life ... Well, at the branch I belonged to, I became their vice-chairman and the whole crew at that time were sure I could assist with problems that ex-servicemen have, which I used to endeavour to do. Sometimes, in the places I've been since—in the care homes and what have you—you find the same problems.

Lakshmi's story spoke of a desire, and perceived competence, to have continued to teach children, and she had only been prevented from doing so by failing eyesight. And while Kannamma realised that her current circumstances were not ideal for teaching, she too retained the hope that her aspirations for the future might still come to fruition:

> I want to do tuition but I can't because of this situation (points to leg) ... it is easier in America. Here it is very difficult to do ... I want, but nobody will come. I want go there only. I will teach... they want Tamil pupils there ... they have Tamil people there so I want to teach Tamil and pongal—you know pongal? (she shows me some rice powder designs from her notebook).

While some of the aspirations that were shared with me spoke of an ambition to 'stay me' by maintaining existing valued identities, there was also often an element of wanting to create new roles and new ways of reciprocating that reflected their changing circumstances and circle of contacts—and plans for a different, but still valued, place-in-the-world. For Megan, her newly found identity as an ambassador for Age UK facilitated reciprocity in the form of being a spokesperson for those she perceived as down-trodden—something she described as typical of what it meant to her to be 'me'. She had earlier described such a situation in her recent past;

> I was on this side of the railings and on that side, right up against the railings, was a little boy… and he was crying like billy-o. And his mother was standing over him and screaming at him, telling him to shut up… And nobody took the blindest bit of notice so I turned back … I said "what do you think you are doing? How can you expect the child to stop crying when you scream at him like a banshee? … What have you done to get him into that state?" … I said "it *is* my business and I've got two minds to report you to the police". God! That's me!

And, having just agreed, at the time of the interview, to reflect the work of Age UK from a service-user perspective, she described this opportunity to continue doing what she had always felt was important, as an act of reciprocity;

> As I say, I feel that I'm paying them back for what they've given me.

It remains, however, that such recognition and facilitation of future aspirations was unusual among the narratives. Because of either their physical reliance on carers, or their isolation from wider communities and opportunities for interaction (or indeed both), elders in both samples usually had need to rely on carers for help in reaching what they aspired to. For carers to have supported these aspirations, however, they would have needed to get to know the participants well enough to have understood what was uniquely important to each particular person's spiritual well-being. Where that dialogue was absent (for reasons explored in later chapters) their identity as a person with a future as well as a past had been lost, or least undermined.

I would suggest that this was not necessarily the fault of uncaring staff, however, as I note that a common theme running through the narratives of the Indian elders in particular was that of 'self-containment'—seemingly a sense of duty to take responsibility for one's own life, and not to involve oneself in others' affairs unless asked to. For example, wherever Vadivu referred to how she might help others, it was always prefaced with '*only if they ask*'. This was typical too of Subbulakshmi's responses, as reported by the translator:

> Talking to people [about emotional support and so on] she'll not do, unless approached.

And it was not only the women who were reticent in this respect. As Venkatraman remarked:

> … to not trouble others with our worries will bring happiness too.

This reticence appeared to extend into not seeing it as appropriate to share details of their earlier lives with their current circle of contacts—either carers or fellow residents—which may have made it difficult for carers to have got to know these older people as unique individuals, even had they been interested to do so.

Self-perception as Changing in Terms of Being a Valued Person

General. This particularly significant theme reflects the self-deprecation expressed by some of the participants, but also a recognition by others that they can still be of use, albeit in different ways than before. A sense of 'unfulfilled life' is represented, as too a life-long ambition to 'make a difference', and to continue doing so despite compromised health, or loss as experienced in a number of forms.

Specific. The dynamic nature of social life has already been implied in the earlier reference to the significance of social space and how their own, and other people's, perceptions of the participants as being capable of reciprocity had changed over time. There is, therefore, little point in going over the same ground here, except to reiterate that for some, especially Megan, Eddie and Muthu, their own sense of self-worth as 'giving' people had been a constant in the changing dynamic.

For some, though, it appeared to have been difficult for them to maintain their self-worth and esteem in the circumstances of becoming dependent on others. Self-deprecation was a feature in several of the accounts where, after feeling that they had been previously been valued for what they had been able to contribute to others, some had come to feel that what they could offer was now no longer relevant. Lakshmi, for example, commented on how her capacity to teach was being affected by not being able to keep up with '*so many modern things*'. Similarly, Kannamma talked of the '*changes made in the academic works*' which made her feel less confident about teaching children than she had done in the past. Ken's account contained elements of self-deprecation, but his understanding of change focused more on his own declining ability, than on external circumstances. Muthu, while perceiving himself to be capable of reciprocating if the opportunity had been available, and often expressing a desire to be 'of use', did not always recognise opportunities when they presented themselves. For example, when asked whether he would be interested to take part in the training of eldercare practitioners by offering a service-user perspective, he did not recognise his perspective as a valid form of knowledge, and attributed his inability, in part, to his ageing:

> I'd be embarrassed… embarrassed because with these people I don't have the capacity to give opinion over theirs. I have only so much learning… age will constrict me. I can't do it.

One final point worthy of mention here is the extent to which individuals who had talked to me at length, and positively, about what had made them feel valued as people earlier in their lives—that is, their strengths—tended to change their focus to weaknesses when talking about their desire and opportunities to reciprocate in their current circumstances of dependency. Edna, for example, counteracted all suggestions of how she might want to get involved in helping others with reference to her failing eyesight or weak heart—even though most of the opportunities would not have involved moving from her chair. Vadivu, too, talked about how she could no longer do anything, even though she *was* doing something to help others, in the form of prayer leading.

Expectations of Old Age in India and the UK: Continuity and Change

General. The equating of old age with a sickness model featured in both sets of responses and has been incorporated in what appears to be a very significant theme because it remains a powerful influence on whether people perceive themselves, and are perceived by others, as capable and competent. Changing attitudes relating to respect for older people in general are also incorporated, as too are perceptions of the particular attitudes of caregivers to competence in old age. The articulation of ageism and sexism was represented in both sets of responses and is included here because patriarchy was perceived by some, though not all, as still pervasive. As is discussed later, this is an example of the double-dialectic described earlier, in action.

Specific. This far, I have made reference to temporal aspects relating to individual experience but, as the participants in the study were not living in a social vacuum, changes relating to expectations *of* them are also significant to an understanding of what can impede or promote reciprocity in their situations. In this respect, the narratives of both sets of participants reflected something of current and changing attitudes towards old age in both settings.

For example, more than one of the Indian elders felt that, in their experience, elders are still revered in Indian society. Lakshmi, according to the translator, commented that:

…even though old age comes, the people here respect old age and they take care of old people.

This view was supported by Vadivu, although she qualified it by commenting that, in terms of respect for women in old age:

'no, there's not much'.

If such reverence still exists, or is perceived as existing, this may have some effect on whether older people are expected to continue to give back to their peers and communities. So too may cultural expectations relating to 'self-containment' as referred to in point 2.1, whereby, if old age itself is not generally understood to necessitate lifestyle changes in terms of self-responsibility to live a useful life, then the duty to help others where one can will retain currency. I found no evidence in the narratives of the seven Indian elders to suggest that they saw old age itself as a reason to stop helping others, or to see themselves as different from their earlier selves, except where old age was accompanied by illness, frailty or disability. Even then, attempts were often made to work around those deficits.

Of further relevance to the lives of the Indian elders in this study, given that they were all Hindus, may have been the expectation of the devout that they should withdraw from the secular world (to a greater or lesser extent), in order to focus on personal spiritual growth. It is difficult to know the extent to which this expectation is observed in the general population of older people in India but, in this study, the religious observances of Vadivu and Venkatraman suggests that this expectation may still have some currency. Jacobs (2010) refers to the role that beliefs relating to duty are said to play in the maintenance of social order:

> ... *dharma* is concerned with the maintenance of the social order. This social order is linked both to personal behaviour and to the maintenance of the cosmos. In other words, the individual, the social and the cosmological are inseparable. The correct behaviour of the individual is perceived as bringing balance and order to both society and the cosmos (p. 57).

In light of this, the process of withdrawal from connectedness with others could itself be seen as an act of giving back—perhaps not to peers, community or even society, but to the cosmos as they conceptualised it?

By contrast, the responses of the UK participants largely reflected an acquiescence to the prevailing negative stereotype of old age as a time of diminishing usefulness to their families, peers and society. While Megan and Ken described challenges to this stereotype—Megan in ways already reported, and Ken, in insisting on having a voice heard in the matter of how resources are rationed to the detriment of older people's dignity—such challenges were in the minority. I could find no evidence in the remaining narratives of a resistance to the prevailing social expectation that older people who need a significant amount of help should not reciprocate, or be valued for that reciprocation.

Intergenerationality

General. Though not mentioned frequently, the potential for mentorship and educational support from older to younger people held significance for at least some of the participants, although not all perceived themselves as competent to offer this. There was also some recognition of generational differences in general, and the continuing relevance of knowledge-sharing across generations despite the concerns of Chadha (2003) that the status of Indian elders as repositories and transmittors of wisdom is threatened by the decline in multi-generational families.

Given that most of the participants described having been able to contribute earlier in their lives to the well-being of children and young people—either as parents or teachers—and having derived self-esteem from that, I had wondered whether they would still regard themselves as repositories of wisdom in their current dependent circumstances. And, if this were the case, I also wondered whether it would have been recognised by care providers and facilitators so, that there might be a two-way exchange of information between generations working together in the co-production of new insights (Hatton-Yeo 2006). However, in considering what impedes or promotes reciprocity in the care of dependent older people, it seems that intergenerational exchange cannot be taken for granted because its promotion or otherwise is somewhat reliant on whether older people are conceptualised as competent to engage productively with, and therefore enrich the lives of, generations to follow them. From the findings of this study, for the most part they appeared not to be.

Specific. There was some, albeit small, degree of evidence in the narratives to suggest that the benefits of intergenerational exchange to the older people themselves had been experienced in meaningful ways. As already described, several of the Indian participants 'gave instruction', as they put it, to the young students who visited Residential Home A, and reported feeling good about being able to do that. However, this had been limited to help with language translation, and in response to approaches from anxious students, rather than as a result of any formal recognition on the part of eldercare practitioners that the participants may have craved recognition of their expertise, or that their well-being could have been enhanced by being recognised as of value to younger generations. As Muthu commented;

> Younger people will not be here. Definitely, they are not coming here … I could give them advice, but no opportunity here… if they come I will do, but there is not the opportunity. Nobody is coming.

And, while by far the majority of the UK elders (five of seven) answered that they would have welcomed, given the opportunity, the chance to work with children and young people, that opportunity had not been facilitated.

However, one account of intergenerational exchange did speak of mutual benefit. Ken made reference to how his grandson had moved on from being his carer, to taking up paid work in the caring profession. Ken appeared to derive some pleasure from having been able to give him that opportunity to learn about caring, and the confidence to make changes to his life. While this may not have been an intentional act of 'giving', it had been recognised as such by Ken, when asked to reflect on reciprocity.

And so, while there had been some instances of informal and self-generated dialogue between generations, there was no evidence of any formal intergenerational initiatives which might have contributed towards the participants seeing themselves as a resource rather than a burden.

Developments in Communication Technology: Opportunity and Barriers

General. Though not overtly mentioned, this was implicit in enough of the responses to constitute a theme which incorporated a lack of opportunity to engage with the potential for technological change to open up new forms and arenas for operating in 'useful' ways—particularly evident in the case of the Indian participants. This was also reflected in the narratives of those in the UK where access to technology was limited to helping with 'administrative-type' activities.

Of consequence to the perception of opportunities for reciprocity is access to the information technology that has revolutionised the concept of social space during the life span of the participants. This is the case for at least three reasons relating to the potential to:

1. open up new opportunities for dependent older people to 'give back' in ways that those with compromised health, strength or mobility might be able to manage *if* they had access to equipment and instruction;
2. help those with existing roles to continue with them, albeit perhaps in different ways, when they can no longer physically access the environment in which that giving back took place; and
3. enable people isolated from their original communities to re-connect with them, or with new ones, thereby increasing their opportunities to build up social capital and the opportunity that can give for the creating of a positive self-image.

However, there was little evidence that this potential was being realised in either the Indian or UK settings. No-one from the Indian sample made reference to having had, or even wishing that they had, access to Internet technology. From the UK sample, only Megan possessed a computer and even she did not use it in a way that could have broadened her social network or enabled her to help others in diverse and imaginative ways. And, from the responses of the four participants who lived in residential care, the only reference to technology was to nervousness about using the office equipment on the occasions where they had been invited to 'help out' in the office. As I discuss in later chapters, that such easily facilitated access (at least in the UK) to a vast array of potentially life-affirming opportunities for social engagement and reciprocity was not reflected in the narratives, invites consideration of why access to information technology is readily available in schools but not, apparently, in residential or nursing homes.

Meaning Making at the Level of the Personal/Spiritual

Have I Got Anything to Give?

General. Concerns of this nature featured very significantly, relating to the variety of ways in which people felt they could reciprocate, whether they perceived themselves to have strengths as well as weaknesses and, on a more negative note, references to unworthiness.

Specific. I have talked this far about the desire to give as well as receive, and of opportunities to do so, but here I focus more specifically on the extent to which the participants saw themselves as competent people—whether reciprocity remained part of their value base, such that they perceived it as appropriate to be seeking out those opportunities in their changed circumstances. The responses fell into three groups:

- Several of the participants felt that they did not have anything to give, and tended to justify that by reference to being ill or 'just old'—perhaps trying to retain some sense of control or mastery by claiming that it was not their fault that they were unable to reciprocate, but that this was not a problem for them. Doris and Edna, for example, tended to frame their response in terms of not being well enough to interact with others or look for ways to help. As Edna said of her doctor's advice to her family:

she has to be careful.

Vadivu also referred to her frailty in response to questioning about what she was engaging with that contributed to her self-esteem at the time of the interview. This was translated as:

> So they never asked her to do those things… and even she thinks she cannot do.

Maureen appeared to attribute what she perceived as uselessness:

> I don't feel worthwhile as a person really …

to old age, in that her self-esteem had always been tied up in having a job (where she appeared to perceive her 'usefulness' to lie), and she claimed to have given up on that aspiration since moving into residential care because she perceived herself to be too old to work.

- Others, including Kannamma, Harold and Megan, seemed more confident that they could be of use to others in some way, but were unsure about whether their input would be valued, or considered outdated in a changing world. For example, knowing that she had formerly been proud of her ability to cook and keep house, I had asked Megan whether she had ever thought of passing on her expertise to others but she replied:

> No, I don't think so. You see, we're a different generation. Kids today don't cook and they don't do anything really. You know what I mean?—you'd be flogging a dead horse, really.

But even if they did not feel particularly valued as people, they seemed accepting of their reduced status, and did not express anger or resentment in their responses.

- Ken and Muthu, however, *did* express strong feelings about having the capacity, and the desire, to reciprocate but perceived themselves as being prevented from doing so. For example, Ken appeared to take it as an affront when his perspective on care provision and dynamics was not sought or valued.

A Sense of Being Valued Rather than Being a Burden

I have already made reference to;

1. the need for affirmation as a good person if one is to feel valued;
2. how this affirmation is only obtainable if one is able to operate in social spaces where feedback from others is available; and
3. how difficult it can be to get positive feedback about competence and worth if one is living out one's life in situations where significant dependence on others is the norm, and highly visible.

However, there was some additional evidence in the narratives, in the several references to teamwork, of a challenge to the assumption that the role of eldercare practitioners is to look after dependent older people, rather than assist them in living the life of their choosing. Where participants perceived the relationship between their carers and themselves as such, it suggested that they may have been trying to make sense of the limitations they were facing by 'reframing' the relationship as co-dependency. It may be significant that, where reference was made to this co-dependency as teamwork, it was the participants who described it as such. For example, Venkatraman said:

> everyone is co-operating, we are well-suited, like family' and added that this 'made me feel good and proud of me'.

Muthu also seemed to take solace from the fact that he could still help others as well as be helped by them—also likening the reciprocity to the give and take that often characterises family life:

> On spiritual times, these people [fellow residents] will help, giving co-operation. We are friends—more than—a family. They come from different families but here, one family.

And Eddie had referred more specifically to how teamwork, as evident in the relationship of mutual support between himself, and his care staff referred to earlier, had contributed to his image of himself as someone who was not a burden on others, especially by saying that he would be *annoyed* had he not been able to have that relationship of reciprocity.

While the men's talk had predominantly been of advice, and the often used Indian-English term 'sedition', for many of the women the challenge had been addressed in a more task-centred way, and often informed by gendered expectations around domestic duties. This had been evident in Subbulakshmi's account in particular, in that she reflected an internalised assumption that her role as a woman of her class and caste was to look after not only her immediate environment, but that of her peers and carers also. This also appears to have been evident in Vadivu's attempts to make the residential home environment more suitable for the worship rituals fundamental to their Hindu lifestyle:

> There is one person here ... she is teacher and we have discussion in reciting and leading prayer. We sit down and talk about these things ... to make this place like an ashram.

The results were therefore gendered, in that they reflected stereotypical obligations which had implications for their sense of selves as not just older people, but as men and women in old age.

Dependency as Existential Crisis: Am I the Same Person or Someone Different Now that I Need Help?

General. This very significant theme reflects the reporting, for example, of a lack of attention to people's past identities or who they wanted to be in the future—a perceived lack of awareness on the part of carers of the 'whole' person. It reflects also reference to risk-taking individuals being denied the opportunity to continue managing risk in their lives.

Specific. From the participants' perspective it is likely that they would have considered themselves to be a 'work in progress', given that they had lived with themselves all of their lives, and therefore knew what was important to them and their own well-being. This had been particularly evident in Megan's story but others, including Eddie and Ken, also reflected a set of core values that seemed still to be to influencing them to continue making a difference by giving back to family, peers or society, even in their current circumstances of being cared for themselves. All of the participants had accepted that they had become physically compromised, but not all accepted that this legitimised their being seen as 'less than'—either in the sense of being of lesser worth than they had been earlier in their lives, or of lesser worth than other people in society.

Venkatraman, for example, seemed to be very happy in his own skin, as the saying goes—focusing on what he perceived to be a positive new phase to his life—as too did Vadivu, who suggested that, while she had deteriorated physically, she remained the same person spiritually, and was very content that she could continue to play a useful role for her peers by facilitating their spiritual duties as practising Hindus. And, as has been referred to already, Megan was proud of being able to help disadvantaged others, appearing determined to maintain the identity she felt had been forged in her childhood, and which had sustained her sense of self-worth ever since.

However, more commonly reflected was disappointment at not being able to reciprocate for whatever reason. Eddie and Harold, for example, had faced serious challenges to their sense of identity as competent people. The disjuncture between pride in their achievements before and since becoming dependent was particularly evident in Harold's case, in that, after a life of service to others, he had come to perceive himself as:

Just someone to be looked after, that's all.

Maureen's narrative also reflected having had to face a challenge to what she had held dear earlier in her life. While she had not shown quite as much confidence in her earlier sense of self-worth as had some of the other participants, she nevertheless seemed to have had some sense of who she was, and what she wanted in life—a job and her own place to live, which would have allowed her to feel engaged with the world:

I thought it was smashing to have a job ... on top of earning money, you're keeping yourself and you're going out and you're *living* [her emphasis] ... keeping my self-respect really.

However, after becoming dependent to the extent that she had been admitted to residential care, her story came to reflect an inner struggle between wanting to regain her former sense of who she had been:

Sometimes I'd like a job but I'm too old... and at other times I think I'd like a little place of my own. I'd like a place of my own—back to myself again.

and giving up on it, to the extent that she appeared to be disempowering herself by acquiescing to a new, and less valued, status:

But now I'm too old so I just let it go by ... just go for my walk every morning, just play by the rules, really.

Perhaps one of the most significant challenges reflected in the findings across both sets of participants, and both genders, was the threat that risk-management policies had posed to the participants' perceptions of themselves as competent, despite needing some help. Being faced with the message from care facilitators and providers to the effect that 'you are no longer capable of making decisions for yourself about reciprocating, so *we* will do the risk management, even though it has been part of who you are for your adult lives up to this point'—proved to be a bitter pill for most, though not all, to swallow.

Giving: Duty or Pleasure?

General. This theme was very significant, reflecting as it did, an understanding of the concept of duty as it relates to a moral code. It incorporates, too, that reciprocity was perceived by some as giving pleasure or benefit to the giver, as well as to the recipient, and that paying for care may have had implications for an obligation to reciprocate.

Reflecting in advance on how the participants may have been trying to make sense of the changes that had occurred in their lives and how these were impacting on their sense of self and place-in-the-world, I had wondered whether the concept of duty had been a driving force for any, or all, of them. It seems that for majority, though not all, it did play a part.

There was no evidence in Maureen's story of her deriving any spiritual benefit from giving to others. She had stated that it had not been important to her, and so it is unlikely that duty to others, apart from deferring to mental health professionals, had been a core value in her life. Similarly, I could not see any evidence that Doris derived pleasure from helping out fellow residents by offering a listening ear, or was driven by any compunction to play that role, other than to relieve her own boredom. That Doris was a paying customer of the care providers, rather than receiving it free as a welfare right, may have been significant to whether she perceived reciprocity as an obligation, a point to which I return in Chap. 6.

Of the remaining participants, however, all agreed that being able to give to others made a positive contribution to their well-being, though not all because of a sense of duty. Edna, for example, had said that she tried to cheer people up because she derived pleasure from that, and because it would 'break her heart' if she were not able to do it. As discussed in point 3. 3, it seems it would have challenged her perception of herself as a gregarious person to whom the well-being of others was as important as her own. Vadivu also appeared to have been driven by the sense of satisfaction that reciprocity engendered:

> She says she feels proud because if she is able to help someway or other, she will be proud of herself—feels happy, feels good.

The majority of the narratives, however, did reflect that the participants' acts of reciprocity were, in part, underpinned by a sense of duty. For some this was reflected as a sense of moral duty. Megan described it as compassion, something that had been an embedded value since her childhood days. She had recalled how the financial hardships her mother had faced after being widowed, and the death of her brother, had changed her as a young teenager to the extent that she felt it necessary to always challenge injustice where she saw it. She had not used the word 'duty' itself, but it was implied in the extent to which challenging injustice was, from then on, a recurring theme in her story.

Eddie, too, while also not articulating it as a moral duty, also returned often to the theme of pride in being able to contribute to the well-being of others throughout his life—in his earlier contributions to national security, the caring professions and left-wing politics, and to the well-being of ex-military comrades in his current situation. At the end of the interview, Eddie had professed a desire to resume attendance at the church that had played an important role in his childhood and adolescence, which suggests that the moral duty he appeared to reflect may have been underpinned by Christian moral values, though this was not overtly expressed.

For some of the Indian elders, reciprocity as a religious duty was implicit in how they lived their lives according to Hindu values such as benevolence and the promoting of harmony, but in Muthu's narrative this was more overtly expressed as a moral duty underpinned by religious values. In discussing ways in which he had given to others, I had phrased the question in terms of what had made him feel proud or special. In response, he replied:

> Ah, not special at all—all men is equal.

adding that:

> God has created me in that light, that temperament—to help others.

However, given that he had also said, in response to being asked why helping others made him feel happy:

> So to go to paradise.

there may have been an element of self-interest in his project to procure opportunities for reciprocity in what he perceived to be his current incarnation. Again, this is an issue to which I return in Chap. 6.

Meaning Making at the Level of Shared Meanings and Institutionalised Patterns of Power

In considering impediments to reciprocity at this level of analysis, I had not expected the participants to articulate their experiences and perceptions using the language of, for example, sociology, philosophy or social theory. Therefore, I had anticipated that rich data of this nature would not emerge as readily as that relating to personal experience and cultural understandings had already done. And, in choosing interviews as a means of producing data, I had taken into account that such a methodological tool had always been more likely to produce insights relating to the impeding or promoting of reciprocity at the personal and cultural levels of analysis, than to structural factors. Nevertheless, while this did prove to be the case, latent thematic analysis highlighted evidence in some of the narratives of an awareness of broader issues than the participants' own concerns, and it is to these that I now turn before moving on to explore structural factors more fully in later chapters.

Health Discourse: Old Age as Illness

General. Examples of what contributed to making this a significant theme include the extent to which care was perceived as holistic; conflict between risk-management policies and personal choice; perceptions of care delivery as dependency enhancing or overprotective; and also the perception by some of their old age as being characterised by decrepitude and therefore only a time for resting and being looked after.

Specific. There was some evidence of reciprocity being impeded by the internalisation of discourses that equate old age with ill-health. For example, some of the participants in both settings had justified their lack of reciprocity by reference to being ill. What was also reflected is that some had been influenced by discourses relating to epistemology to assume that, as 'sick' people, their opinions had less legitimacy than those of health or social care staff. Muthu, for example, commented that, even if he had been given the opportunity to contribute to the training of health and social care staff, he would not have:

> ... the capacity to give opinion over theirs.

The internalisation of ageist and medical discourses, and the implications for their experience relating to the impeding or promoting of reciprocity, was reflected in the fact that, though many of the participants had become knowledgeable about their chronic medical conditions through personal experience, none had been asked to use this for the benefit of newly diagnosed or less-informed peers, nor expressed the view that it was odd that their perspective had never been sought. In a similar vein, Ken had had a lot to say about what he perceived as wasteful resources management in the delivery of eldercare:

> When I go to the hospice. I go to t'toilet, and when they give me a piece of, you know, what they might have—wiping paper—they don't only pull one piece, they're pulling it off and pulling it off. You've got a big bundle in your hands. Now, to me that's disgraceful and I tell them that...

but there was no evidence to suggest that his opinions had been sought, or listened to, by those with the power to make changes, or that he had been surprised by this.

Social Construction of Old Age

General. While not featuring explicitly, as was only to be expected, inferences to it made this a very significant theme. Reflected were understandings of respect (or lack of respect) for old age; the marginalising of dependent older people from decision-making processes; and assumptions about older people's capacity to learn. A broad theme, it also encompasses examples of eldercare focusing on keeping people active, rather than on helping then to continuing to develop as individuals, and also the articulation of sexism and ageism.

Specific. In addition to earlier points about the respect, or otherwise, accorded older people in India and the UK, there was some evidence of the participants being conceptualised as a homogeneous group, an assumption typical of ageist ideologies which serve to make it seem unremarkable that older people are treated as such, especially when they live in group settings (Nay 1998; Knight and Mellor 2007). There was evidence of this in both the UK and Indian samples, in both instances it being in respect of it being seen as appropriate to provide 'activities'—sessions where the participants were expected to join in activities that had been chosen by others on the assumption that these would be something that 'old people' would like to do, regardless of differences relating to interest or skills levels.

For example, Ken found himself making trinkets at the hospice—an activity way below his level of technical skill. Similarly, Maureen and Edna were putting plants in pots (though one had shown little interest in gardening and the other already had expertise beyond potting up seedlings) and making fairy cakes from packet mixes which, to my mind, may have seemed quite patronising, at least to Doris. Even in Chennai, in Residential Home A, where 'self-containment' and individual responsibility in line with ability seem to have been normative values, it had been seen as acceptable to organise a joint activity session (flower arranging) without reference to a) whether or not it would have been of interest to the residents or b) whether anyone had the expertise to play a part in *demonstrating* flower arranging skills, rather than being assumed to be in need of being taught. Megan's narrative speaks particularly poignantly of how it can feel to be denied one's individuality on the grounds of age alone. When considering a move to a housing facility for older people, she had been told by an associate that she would be bound to like it because there was a 'lovely bingo club there'. This was her response:

> Aaaah, God! I think 'Well, how do I stay out of that one then! I want to do my own thing. I don't want to do other people's things, you know.

Discourses Around Dependency: Care Model as Overprotective

General. This very significant theme arose from understandings of a focus in care provision on deficits, to the detriment of strengths, and the labeling as dependent by others in revered or authoritative positions. Also categorised under this theme were reflections on, or references to, the relationship between poverty and reciprocity and also how the concept of 'helping' was understood.

Specific. The narratives of the Indian elders tended to support the conceptualising of formal care as providing a relatively safe environment in which they could carry on their lives as independently as possible, calling on help only when needed—as, for example, when Lakshmi or Vadivu wanted help to visit the temple in the grounds. In such an environment, the participants described being able to help each other in ways that would have been fairly typical of interactions with friends and family members.

The UK-based participants spoke more of a 'managed' environment, where care appeared to be conceptualised more as protection from risk, than as help to live with risk—a point developed further in later chapters. While some of the UK elders regarded this as challenging to their perception of themselves as competent people (Eddie, Harold and Ken, for example) few of them resisted by taking the risks anyway. Both Edna and Maureen appear to have been particularly heavily influenced by the assumption that they should devolve decisions about risk to 'those who know better—Maureen accepting the judgement of mental health professionals that she was not competent to make decisions for herself, and Edna accepting her doctor's judgement that she had to be cared for, rather than helped to live the life she wanted to live within her limitations.

Furthermore, there was very little evidence in either sample, of anyone focusing on the participants' strengths (Saleeby 2008)—that is, the ways in which they could help others—rather than their weaknesses, for which they needed to receive help. The exceptions to this were Megan's and Eddie's narratives, but considerations other than the possibility that their care providers were committed to an empowering care model may have been informing their experiences, as is discussed further in Chap. 6.

Welfare Policy and Practices

General. Of some significance, and therefore constituting a theme in themselves, were inferences to systems-led cultures in residential homes and domiciliary care provision, to the detriment of the unique ways in which reciprocity could be facilitated, and also to the effectiveness or otherwise of service user initiatives and the opportunities they presented for 'giving back'.

Specific. I was not expecting much evidence in the findings that would relate to an understanding of matters relating to welfare policy and practices at a level beyond the cultures of the individual residential homes and care providers, and this proved to be the case. The following point is worthy of mention, however, and its implications are discussed in greater detail in Chap. 6.

Despite national policy in India espousing old age as a time for 'contribution to society' (National Policy for Older Persons 1999), rather than one typified by dependency and welfarisation, there was no evidence to suggest that the participants in this study perceived this as being promoted in their personal circumstances. None, for example, reported being enabled to engage with their local communities, where opportunities to have fulfilled that promise might have been forthcoming. And while current UK policy on ageing may also be promoting more involvement of dependent older people in decision-making processes (the Better Government for Older People initiative, for example), Megan's was the only narrative that provided any evidence that would provide a serious challenge to the stereotype of older people as a burden rather than a resource.

Conclusion

Limitations of space have required me to be selective about the findings I have been able to present in this chapter, but those I have chosen reflect that opportunities for, and impediments to, reciprocity related not only to personal beliefs and attitudes, but also to differences in social structure, and to prevailing dominant ideologies in the societies in which the eldercare was experienced.

In response to my research questions—in what ways are social space, social time and meaning making significant for an understanding of reciprocity in the lives of dependent older people?—the narratives provide evidence to suggest that, across both groups and for the most part:

- their aspirations to continue in previously valued roles, or engage in new ones, often went unrecognised and, as a consequence, were often not addressed by those providing or arranging their care;
- they wanted social affirmation because they connected it with spiritual wellbeing, but did not often get the opportunity to operate in environments where intersubjective feedback would have been made that social affirmation available;
- they felt that they were currently being treated differently, and with less regard for personhood, than they had been previously in their lives—although the extent to which this constituted an existential crisis differed both within, and between, the two groups; and
- their accounts reflected how their meaning making had been influenced by differing ideological constructions of old age, the normative rules associated with them in the two different cultural contexts of India and the UK, and the discourses which arise from, and sustain them.

As such, then, the impediments to reciprocity were not only practical, but also ideological. In almost every case, the participants in both cultures, *wanted* to reciprocate in some way as a counterpart to their dependency, but encountered barriers which relate to a number of levels of analysis, and it is to these that I turn in Part Three.

References

Becker, C. S. (1992). *Living and relating: An introduction to phenomenology*. London: Sage.
Bond, J., & Cabrero, G. R. (2007). Health and dependency in later life. In J. Bond et al. (Eds.), *Ageing in society: European perspectives on gerontology*. London: Sage Publications.
Bond, J., Peace, S., Dittman-Kohli, F., & Westerhof, G. (Eds.), (2007). *Ageing in society: European perspectives on gerontology* (3rd ed.). London: Sage.
Bytheway, B., Bacigalupo, V., Bornat, J., Johnson, J., & Spurr, S. (Eds.), (2002). *Understanding care, welfare and community*. London: Routledge.

References

Chadha, N. K. (2003). What motivates intergenerational practices in India? *Journal of Intergenerational Relationships, 1*(1), 177–178.

Grenier, A., & Hanley, J. (2007). Older women and "frailty": Aged, gendered and embodied resistance. *Current Sociology, 55*, 211–228.

Harrington Meyer, M. (2000). *Care work: Gender, class and the welfare state*. London: Routledge.

Hatton-Yeo, A. (Ed.), (2006). *Intergenerational programmes: An introduction and examples of practice*. Stoke-on-Trent: Beth Johnson Foundation.

Hickey, G., & Kipping, C. (1998). Exploring the concept of user involvement in mental health through a participation continuum. *Journal of Clinical Nursing, 7*, 83–88.

Hudson, J. (2002). Community care in the information age. In Bytheway et al. (Ed.), *Understanding care, welfare and community: A reader*. London: Routledge.

Jacobs, S. (2010). *Hinduism today*. London: Continuum.

Knight, T., & Mellor, D. (2007). Social inclusion of older adults in care: Is it just a question of providing activities? *International Journal of Qualitative Studies in Health and Well-being, 2*(2), 74–85.

Nay, R. (1998). Contradictions between perceptions and practices of caring in long-term care of elderly people. *Journal of Clinical Nursing, 7*, 401–408.

Pascal, J. (2006). *The lived experience of cancer survivors: Heideggerian perspectives*. PhD thesis, LaTrobe University, Bendigo.

Saleeby, D. (2008). *The strengths perspective in social work practice* (5th ed.), Rugby: Pearson Education.

Thompson, N. (2000). *Theory and practice in human services*. Buckingham: Open University Press.

Ungerson, C. (2000). 'Cash in care'. In Harrington Meyer (Ed.), *Care work: Gender, class and the welfare state*. London: Routledge.

Part III
The Implications

Chapter 6
The Significance of the Findings for the Spiritual Well-Being of Older People Dependent on Formal Care

Introduction

In this chapter, I discuss the significance of the findings in relation to how the participants' narratives augment or challenge the existing knowledge base relating to reciprocity and well-being, and the extent to which the findings endorse the value of a phenomenological focus.

Broadly speaking, the findings have indicated that:

(1) reciprocity had been, and remains, important to the participants; and
(2) is perceived by them as not being recognised by those in a position to support them in maintaining a valued identity from which they can derive self-esteem.

More particularly, the significance of these findings relates to:

- loss of personhood—how neglecting the need for social affirmation has the potential to depersonalise older people by contributing to the processes of medicalisation and welfarisation that can mask their uniqueness as individuals *with* problems, rather than *as* problems;
- epistemological concerns about the value accorded older people's perspectives;
- how the conceptualisation of reciprocity and spiritual well-being is both context dependent and fluid and,
- more particularly, the potential of the findings to provide a challenge to the conceptualisation of older people who are dependent on formal care as having a past and a present, but no future.

I therefore highlight that the personal narratives reflecting these outcomes are embedded within cultural and structural contexts—shared understandings of old age and eldercare at the cultural level (Chatterjee et al. 2008; Gilleard and Higgs 2010) and hierarchical social division at the structural level (Thompson 2005; Timonen 2008; Jacobs 2010). Given that the focus of the research has been on the perceptions of the participants (at the personal level), who were experiencing dependent old age in differing contexts and informed by differing discourses (at the cultural level), it may appear that only the dialectical relationship between those two levels of analysis is of significance.

However, as I have been influenced by the PCS framework presented in Chap. 2, I draw also on Thompson's reference to a second dialectical relationship—that

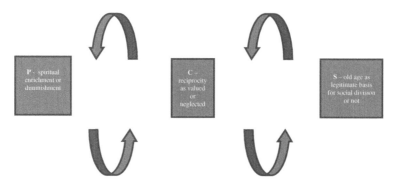

Fig. 6.1 Shared cultural assumption and the double dialectic

between shared understandings of old age and eldercare (at the cultural level) and the way in which societies are categorised into groups imbued with social significance, including age (at the structural level)—to further strengthen the challenge to atomistic approaches which fail to take relationships of power into account (Rubenstein 2001).

Below, I revisit the diagrammatic representation of this double dialectic that can be seen to be operating to impede reciprocity in some contexts, and promote it in others (Fig. 6.1).

In the discussion that follows I make reference to how this framework can help to account for the similarities and differences evident in the narratives of both sets of participants in relation to their experience of promoting, or impeding, of reciprocity. In doing so, I demonstrate how the dialectical processes it proposes can be seen to operate in the same way in both the UK and India, but reflect differences relating to cultural understandings and social structures in the two settings.

As with Chap. 5, this discussion is structured around those concepts already established in Chap. 2 as significant in social life—social space, social time and meaning making—and the themes used to structure the discussion incorporate those which emerged directly from the latent thematic analysis of the transcripts.

Section One

In response to the first of my research questions:

In what ways is the concept of social space significant in accounting for how reciprocity in the care of dependent older people in receipt of formal care is promoted or impeded?

three themes emerged from the literature review, and the empirical research, which suggest that reciprocity is likely to be impeded (or, conversely, promoted if these conditions exist) unless:

- there is a recognition that relationship building and maintaining are as important in dependent old age as they are at any other stage in the life course;

- opportunities to 'give back' or remain 'useful' when in receipt of formal care are recognised as such, and facilitated; and
- barriers, both physical and ideological, are recognised and addressed.

All three themes have significance for the sense of personhood that can be undermined where the spiritual need for affirmation is neglected. It is to the first of these that I now turn.

Building and Maintaining Relationships

It has been established in the existing literature that individuals need to operate in social spaces that constitute, or facilitate, relationships with others in order to be able to:

(1) situate themselves in the world in order to feel socially validated (Crossley 1996; Attig 2011) and
(2) build up social capital through mutually beneficial relationships, so that they have a resource on which they can call in times of adversity (Bourdieu 1986; Putnam 2000; Lin 2008; Gray 2009).

But in order to maintain their identity as partners in reciprocity, and not just benefit from other people's largesse, they also need to build and maintain relationships so that they have opportunities to put something into their own social capital 'bank' to bolster their own sense of spiritual well-being, and in order not to feel diminished by drawing from it when they need help themselves. In highlighting that having a valued 'place-in-the-world' remained important to the participants, the findings largely support the premise that being able to operate in social environments in which they feel valued is significant for older people in terms of positive self-esteem (Victor et al. 2009), and that being seen by others as someone willing, and competent, to reciprocate remains important to their well-being (Moriarty and Butt 2004; McMunn et al. 2009), despite, and maybe because of, increasing dependency on other people. This reference to 'feeling valued' is crucial in terms of this study because it helps to differentiate between social engagement in general, and reciprocity as a particular element of that. It is crucial because my discussion of reciprocity is concerned (a) with older people's perception of social engagement as life-enhancing because of the potential it holds for making them feel valued and (b) how restrictions on social engagement both reflect, and are reflected by, a lack of opportunities for reciprocity.

As highlighted in Chap. 5, all of the participants were, albeit with prompting, able to identify contexts in which they had previously operated as people with something to offer others in the various social contexts and relationships in which they had lived throughout the course of their lives. For some, the prime source of feedback which had made them feel socially validated had come from the public sphere of work, but pride in competence within the domestic sphere had also been represented in several of the participants' stories. Given the age cohort, and the social expectations operating at the time when those I interviewed would have been wage earning and child rearing, I had anticipated that the men might have located their sources of validation primarily in the public sphere, and the women primarily in the private sphere, but this did not prove to be the case. While several of the men had described feeling good about their competence and 'usefulness' in their workplaces earlier in their lives (Eddie and Ken, for example) others, including Venkatraman, Muthu and Harold had been somewhat self-deprecating about what they had been able to offer there, and drew more readily on supporting family members in difficulty, or their voluntary services to the community, as examples of relationships that had been significant in terms of their self-perception as valued individuals.

Where the Indian elders were concerned, this may partly be explained by reference to the dialectical relationship between the individuals looking for affirmation as a 'helpful/useful' person (at the P level), and a shared set of meanings in Hindu-influenced Indian culture (the C level) which informs them that it lies in loyalty and obligation to one's community as part of the broader concept of duty or dharma (Chatterjee 2008; Farndon 2007).

And while some of the women in the study located their validating experiences primarily in the domestic sphere, (Doris and Kannama, for example), for others (Megan, Edna and Lakshmi) their sense of 'usefulness' was reflected as much in their recollections of paid work as it was in their roles as mothers, wives, sisters and daughters in the domestic sphere. That almost all of the women referred to the sphere of paid work in their responses may be partly explained by:

- the articulation between class and gender—in that most of them had little choice but to be wage-earners instead of full-time homemakers. The few women who did locate their opportunities to reciprocate (and their pride in doing so) predominantly in the home may have done so because they were in financially secure households and, as they had not done any paid work outside of it, had only their domestic role to draw on as experience of reciprocity. The ideological assumption dominant at the time when this cohort was younger—that women's role as home makers was taken for granted and unremarkable, (Rowbotham 1973)—might help to explain why they had framed their homemaking as an act of 'giving', when I had asked for examples of reciprocity.
- differing cultural expectations—in that some of the Indian women referred to not expecting to have their contributions respected in old age, or to expect affirmation in general, because they had never had that as women in Indian society anyway. Their low expectations may account, in part, for the acquiescence reflected in their accounts of life in residential care. Where the UK women reflected acquiescence, it was not to the same degree, and appeared to be linked with expectations around age, rather than gender. As such, the implication in the findings that:

(a) for the Indian elders, gender was perceived as having more of a discriminatory impact on their life opportunities than age; and
(b) for the UK elders, age was perceived as having more of a discriminatory impact on their life opportunities than gender,

supports the premise that ageist and patriarchal discourses are fluid, inter-related and embedded in social structure (Arber et al. 2003a; Bagga 2008; Jacobs 2009).

- Cultural differences in how old age are conceptualised in India and the UK. In India, it would seem that expectations of self-sufficiency and self-responsibility are taken to be relevant across the whole life span (Farndon 2007) and, therefore, still relevant in old age. As such, then, it is considered that only severe infirmity in old age would prevent participation in the public sphere, rather than old age *per se*. And so, even if they were not actually still working in the public sphere at the time I interviewed them, it is likely that self-sufficiency and self-responsibility, would still be in their consciousness—part of the mindset by which they were making sense of their current situation of being in receipt of formal care. It was a concept that ran through several of the Indian narratives, specifically in terms of a duty to help others, which appears to reflect that, while 'Hinduism' may have gone through changes in light of the interaction of Hindus with western ideas, its tenets nevertheless continue to influence the daily lives of Hindus (Varma 2005).

This third point might account, in part, for why those caring for the Indian elders may not have seen providing opportunities for reciprocity as their responsibility—or if they did, then not as a priority. The very fact that I had been exploring the issue of being helpful or 'useful' at all, caused some degree of puzzlement to one of the home managers, and several of the participants.

I interpreted this as surprise that what they saw as a taken-for-granted expectation of communal life, and moral and spiritual fulfilment, was of such interest to me. Yet, on reflection, I should not have been surprised, given how embedded those assumptions are, as has been discussed, in the religious and philosophical discourses that have informed their lives.

To many of the Indian elders, the importance of maintaining a valued role through acts of reciprocity appeared to be directly related to a religious duty that was taken for granted, though I would suggest that Muthu's experience of being denied opportunities to help others when he felt a moral imperative to do so supports the plea by Gandhi and Bowers (2008) for issues relating to morality to be higher on the social care agenda. For those in the UK sample, this moral obligation was less evident and may be explained, in part, by having internalised low expectations by others of their competence to reciprocate, (operating at the C level) and were therefore less concerned about the consequences of being defined only as dependent, rather than as multidimensional and competent individuals.

Furthermore, the relatively low evidence in the UK narratives of a moral imperative to reciprocate may relate to:

(a) the more secularised nature of their social context; and
(b) the existence of a shared understanding of the function of the welfare state as being to take responsibility for their well-being, while also recognising that this has changed during their lifetime (Jordan 2008).

This demonstrates how shared understandings at the C level are not set in stone, and the dialectical relationships between the personal, cultural and structural levels fluid (Thompson 2011a). For example, where older people are perceived as worthless, this can be seen to strengthen the legitimacy of a social structure in which age is significant—in that old age is conceptualised as being separate from adulthood, rather than a stage along a continuum of adulthood (Midwinter 1990). But if challenges to that perception of worthlessness are consistent and sustained, such as:

(a) those emanating from legal, academic or user-involvement perspective; and
(b) the highlighting of good practice where aspirations to reciprocity are facilitated, the justification for discrimination can be undermined, and the association between old age and worthlessness weakened. It seems that Age UK workers recognised Megan's potential in this respect, when they urged her to join their cause in the interests of promoting both her own spiritual well-being, and the broader interests of her peers and future cohorts of dependent older people. That neither Megan, nor Muthu, appeared to accept much help from others, but were extremely enthusiastic about *giving* help to others, reflects and supports existing evidence that individuals tend to prefer to under-benefit from relationships of reciprocity (Uehara 1995).

The issue of compromised personhood was reflected in responses that referred to a desire not to be seen as a burden. In resisting the process of 'welfarisation' (Fennell et al. 1988) it could be argued that UK-based elders such as Eddie, Ken and Megan were challenging culturally sanctioned assumptions that to be old in the UK is to be a burden on society—a shared understanding rooted in, and sustaining, a social structure where age is seen as a legitimate basis for discrimination because older people are considered to be not only different from, but 'less than', other adults (de Beauvoir 1977; Thompson 2007; Neuberger 2009). However, we have seen that, in Indian society, age does not currently appear to be *as* significant as a basis for social division because of the primacy of family and interdependency within it (Datta 2008), and older people not regarded as 'less than' others to the same extent, and so they may not feel themselves to be a burden to the same extent. This was reflected in the findings as, while Muthu did explicitly refer to not wanting to be a burden on his family, this was not generally a significant feature of the Indian participants' narratives.

For both groups, reflection on their remembered past and present experiences, has suggested that being in reciprocal relationships where they felt valued was significant for the participants' sense of spiritual well-being because it had facilitated feedback about continuing to be a 'good person'. For example, in Kumari's case, the role she had played providing emotional and practical support to her peers in times of crisis (including death and dying) appeared not only to be:

(a) a source of affirmation and validation in itself –

She will give physical and moral support to us in the critical time. This is her best service to the home.

and also

(b) enabled her to engage with a broader network of people than she might otherwise have done while confined to a care facility. This then held potential for her to build up reserves of social capital—support on which she could draw in times of her own need, and from which she could get intersubjective feedback that could enhance her self-worth based on other people's assumptions that she was a 'good person to know'.

In terms of the promoting or impeding of reciprocity in the care of older people dependent on formal care, this latter point has significance, in that Kumari's story provides evidence of how a visible demonstration of competence (at the P level) by someone assumed not to be competent (at the C level), has the potential to challenge the negative stereotype of older people as incapable of reciprocity. This affirming of competence to a more public audience reflects the reference by Putnam (2000) to 'bridging' social capital, but, more significantly, highlights the potential for a challenge to be mounted in relation to power relations that can function to impede opportunities for reciprocity. In demonstrating that an elderly resident had been capable of undertaking duties more usually undertaken (at least in the UK) by eldercare practitioners, the findings have supported the premise that 'vertical ties' relating to power (Fennema and Tillie 2008) have significance for an understanding of how social capital operates, and, therefore, for a better sociological understanding of the impeding or promoting of opportunities for reciprocity.

Similarly, in Megan's case, her being approached by an Age UK employee to act as an advocate and user representative on their behalf provided an opportunity to bolster her own self-esteem and spiritual well-being through social validation as competent to reciprocate. More than this, however, that it facilitated access to new social networks within which her skills and competence would be visible to others (and which would provide opportunities for a positive association between old age and reciprocity to be reinforced in a public arena) held significance for spiritual well-being beyond her own.

Eddie's and Harold's experiences of being prevented from operating in wider environments which had the potential for social validation, demonstrates the other side of the coin by highlighting a *lack* of opportunity for feedback about being a 'valued person', because opportunities for increasing their access to validating relationships had been closed off, and the social capital that might accrue from them, lost. While operating in the public sphere of work (an accepted norm earlier in their lives) may have facilitated interconnectivity and validation at that time, ageist discourses suggest that, while it may be the norm for men in general, such validation is no longer 'necessary' for the spiritual well-being of *older* men.

So in terms of intersubjective feedback, Eddie and Harold had the perceptions of only those who had no knowledge of them other than as dependent older people, to feed into their own self-image, which may help to account for their frustration, and self-description as worthless, as presented in Chap. 5.

Are Opportunities to 'Give Back' or Remain 'Useful' When in Receipt of Care Recognised as Such and Facilitated?

When I was talking with the participants about how their identities and self-esteem had been constructed in the past, they had all spoken to me of those aspects that had made them feel proud or competent—in essence, what had helped to define them as unique people with unique ways in which they could reciprocate, and so feel useful and valued. One might have expected care providers and care managers to have asked similar questions during assessment or reception processes about what had been, and remained, important to them—especially if they claimed to be addressing holistic needs. However, the findings indicate that only one person (Harold) remembered any such issues being addressed during his needs assessment process, and that it was *he* who had identified a need to share learning experiences, rather than his aspiration to intellectual growth, and being of value to others, having been recognised by the person assessing his needs. In perceiving that he was not being related to as a unique person, with a need for connectedness and the positive affirmation it promised, Harold's experience supports the premise made by Bowers et al. (2001) that, where eldercare is experienced as instrumental, spiritual well-being tends to be compromised.

Furthermore, the findings suggest that, where the participants did find opportunities to operate in social spaces where they could counteract their dependency through reciprocity, these were most often as a result of their own initiative, or in response to invitations from their peers (such as Vadivu's prayer leading and Doris's somewhat reluctant role as 'Agony Aunt'), rather than having been instigated by care facilitators or providers operating within a particular ethos or culture that espoused reciprocity as a core value.

For example, that both Eddie and Maureen (residents of the same residential home) were invited to help out with clerical work, had initially suggested to me that there may have been a recognition by the home's manager of the significance of reciprocity for self-esteem and spiritual well-being, and a commitment to promoting it through a shared organisational culture that recognised the spiritual need for positive affirmation as valued. Though it is possible that such a culture did exist there, in both cases there appear to have been therapeutic issues that might have underpinned those invitations more pertinently—a perceived need for distraction from particular forms of supposedly 'unacceptable' behaviour in Eddie's case, and an attempt to engage Maureen in any activity, regardless of its nature and value, given her tendency to retreat into solitude.

In residential settings in both India and the UK, there had been examples of participants reporting that they had been invited to meet with eldercare practitioners to comment on care provision from their perspective—though this had not necessarily been conceptualised as reciprocity by the participants or, indeed, by the care providers. However, insights provided by Muthu and Doris suggest that these may have been fairly tokenistic attempts at user involvement, or at best, requests for responses to decisions already taken by others—which Hickey and Kipping (1998) placed near the opposite end of the spectrum from 'user control' in their user participation continuum model. It is difficult to establish whether those invitations to 'be of use' were prompted by a genuine understanding of dependent older people as multidimensional, and still competent to give back in some respects, in spite of illness or impairments, but it seems clear from the meaning making inherent in the responses of the participants that, if this were the case, the message that their opinion mattered did not get through to them anyway.

Tokenistic commitment to user involvement (Lloyd 2008) has the potential to operate as an impediment to reciprocity by giving the participants the message that it is taken as read that they have nothing useful to say. Drawing again on PCS analysis, it can be argued that such tokenism on the part of care providers, and the participants' internalising of this feedback that their perspective is not worth inviting, has the potential to reinforce the assumption (at the C level) that dependent older people have less 'value' than other adults, which can itself then reinforce the justification (at the S level) of age as a social division.

Megan's experience of being asked to reciprocate for the support she received, by promoting the aims of the organisation who supported her, provided some evidence to counteract the much more dominant perception in the findings that opportunities to give back had not been recognised or facilitated. This anomalous finding may perhaps be accounted for by the fact that the organisation inviting her to use her talents to heighten their profile (in the interests of helping many more older people) was not a care provider or facilitator, but one whose *raison d'etre* is to challenge mutually reinforcing ageist and medical discourses, and promote more empowering citizenship discourses that espouse rights and personhood. As such, its employees may perhaps have been less likely than those working in the residential care homes and domiciliary care agencies involved in this study, to have been influenced by a 'care' model which does not easily accommodate reciprocity (Bowers et al. 2001; Thompson and Thompson 2001).

It emerged from the narratives that there was a general consensus that carers and care facilitators tended to impede opportunities for reciprocity by failing to recognise, and draw on, their strengths, including expertise in teaching, administration, healthcare, domestic management, gardening, local history and technical know-how. This can be seen to support the premise that strengths-based initiatives (Moss 2005; Saleeby 2008) are not often associated with older people's needs (Lamb et al. 2009) and particularly their spiritual need to feel valued. There had been some evidence in the findings of the participants having been self-deprecating about their expertise and knowledge, in suggesting that it would be outdated and therefore of limited use (an epistemological issue to which I return later). Furthermore, there was also evidence of the perception that they did not have the physical strength to reciprocate, either in ways they had done previously, or in new ways. However, regardless of meaning making relating to the participants' understanding of reciprocity, there was little evidence of any attempt on the part of care facilitators or providers to counteract that acquiescence to dependency by modifying the form in which that expertise could have been drawn on.

As discussed earlier, the opening up of 'virtual space' has had immense implications for the conceptualising of community (Hudson 2002; Hiller and Franz 2004). Yet there was no evidence in the findings that emerged from either setting that the potential for accessing this in the interests of promoting opportunities for connectedness and reciprocity had been recognised or realised. Poverty may have been a factor in this, given the correlation between low income and advanced age in both settings (Wilson 2000; Chakraborti 2004), but perhaps also that dependent older people in general are not recognised as worthy of investment in the necessary equipment (an assumption at the C level), or deemed capable of using it? Where the latter is the case then I can see value in arguments that suggest that this may be changing as new cohorts age with the relevant technical skills and desire to use the technology. If, however, it is an individual's level of dependency, rather than their age, that is used to define them, as may be the case where the Indian participants are concerned, then such technologies may continue to be denied to dependent older people regardless of age cohort, unless ageist assumptions about competence and worth are undermined.

The findings have highlighted that cultural assumptions and norms, and the attitudes and cultures of care staff and service providers, have the potential to operate as potential impediments to reciprocity in the lives of dependent older people.

I now move to consider the particular significance of real, or assumed, boundaries as aspects of social space that may impede the facilitation of reciprocal relationships.

The Recognition of Physical and Ideological Boundaries as Potential Barriers to Reciprocity

The most obvious example of a barrier was that inherent in the living arrangements of the participants, such that they were physically segregated (through ghettoisation or isolation) from the communities that would have provided:

- a sense of connectedness with the wider world and the intersubjective feedback they needed to be able to situate themselves as valued people within it (Crossley 1996; Attig 2011); and also
- a sense of continuity with how they had perceived themselves as operating in that world prior to becoming ill or frail—that is, a sense of their continuing biography (Webster and Whitlock 2003; Tanner 2010).

As already identified in Chap. 4, 11 of the 14 participants in this study lived in residential establishments, among peers whose health and mobility were compromised to the extent that they could not voluntarily engage with their local communities and, of the three who were supported to live in community settings, only one was able to leave home spontaneously, or without considerable assistance. This meant that opportunities to give back or feel of use were, for almost all of the participants, limited to their immediate environments, especially as they did not have access to the wider social environments being opened up by advances in Internet technology.

In their narratives, most of the participants reflected acquiescence to their isolation (even where they regretted it), and it was only in Eddie's case that segregation from the wider community seemed to be actually distressing for him. For others, it may have led to boredom, or a sense of being under utilised, as in Muthu's case, but there was often a sense of inevitability reflected—as in Maureen's comment that she 'just goes by the rules, really'. While it was not expressed as such, the acquiescence of the Indian elders to their isolation from wider communities may have been a reflection of their having had little other choice, given that domiciliary support services were very limited, and available only to those with the means to pay for them. Without kinship care, or an infrastructure of domiciliary care support, and in a context where many vulnerable individuals have to live without shelter on the streets, they may have felt both grateful for what they had, and also vulnerable by being beholden to their benefactors. As a consequence, they may therefore have been more willing to acquiesce to the rules of the care cultures they found themselves living within. While the ethos of self-responsibility referred to earlier may have been a factor in the impeding of reciprocity for the Indian elders, the acquiescence shown by the UK elders perhaps reflects more readily the 'musn't grumble' attitude described by Pickard (2010), as being representative of the 'self-policing' processes described by Foucault (1977) and Gilbert and Powell (2010).

The narratives also suggest that style of accommodation provision may have been a significant factor in relation to the promoting or impeding of reciprocity. Both of the residential homes from which the Chennai-based participants were drawn operated dormitory style living arrangements (with the small temple and garden comprising the only communal area), while those in the UK sample who lived communally each had their own room, with the invitation, but no compunction, to eat and spend time in communal areas. It might have been assumed that those living in Chennai, being always in close proximity to other residents, would have reported more opportunities to connect with others in ways that fostered reciprocity. However, the expectation of, and adherence to, an ethos of self-responsibility operating at the cultural level may have constrained them from indulging in acts of reciprocity, and getting involved in each others' affairs unless expressly invited to.

Furthermore, both of the residential homes in Chennai from which the participants were drawn were located quite some way out of the city centre, and engagement with the community outside of them was only possible for those with the financial means to access them. In the case of this study, only one—Kannamma—had that option. In an attempt to address the ghettoisation of older people in care homes, the residential home from which four of the UK participants were drawn had been modified to create a village-style layout and atmosphere, with residents living in three different living spaces set around a communal garden area. In terms of promoting reciprocity, this may have reflected a recognition that identity is associated with place (Renzenbrink 2004; Peace et al. 2006) and an assumption that a sense of community is beneficial for well-being. Crucially, however, in terms of the potential to foster a sense of continuing personhood through feeling valued, the residential community was nevertheless still segregated from the town in which it was situated, and access to it did not seem to be easily facilitated—seemingly because of a combination of risk-management policies and staff availability, as was highlighted in Eddie's narrative. Eddie's perception of himself as someone competent to have a useful role outside of the home's 'closed' community, but being denied the opportunity to operate there, provides an example of;

(a) a spiritual need (that of wanting to have a role to play in the world and feeling valued because of that) being overridden by the perceived or actual need for protection from physical harm; and

(b) the power of the combination of ageist and bio-medical discourses to make defining dependent older people only by reference to their physical deficits seem unremarkable.

While conducting the interviews, I saw no evidence of lateral or creative thinking on the part of care providers in either country in relation to how the promoting of opportunities for reciprocity could co-exist with the management of risk. The earlier lives of all of the participants in this study had been characterised by risk taking—it had been an integral part of who they had considered themselves to be, and how they had lived their lives before finding themselves dependent on others for help with daily living. Had those providing or organising that help accepted that those who had been competently managing risk previously could continue to manage risk in the recognition that their health had become compromised—and had they challenged risk averse practice (Manthorpe 2007; Bornat and Bytheway 2010) by helping those in their care to manage new sets of risks—then I would argue that most of the affronts (such as those described by Eddie and Ken, who wanted to reciprocate, but had experienced barriers being put in their way) could have been avoided. Furthermore, their spiritual well-being would not have been under threat of being diminished by reifying processes that portrayed them not as individuals living with risk, but as risks to be managed themselves.

It is possible that a less segregated environment than a care home might have facilitated the breaking down of ideological boundaries between youth and old age, through the potential for intergenerational initiatives to challenge negative stereotypes of both older and younger people (Thompson 2005; Hatton-Yeo 2006). Given that the 14 participants had included teachers, a doctor, a school matron and a librarian, the potential for mutually supportive and spiritually enriching interactions with younger people seems to have been significant, but overlooked. Again, a missed opportunity to challenge negative assumptions held at the cultural level of shared meanings (which ultimately have an influence at the structural level on how dependent older people live, or are allowed to live, their lives) can be explained by operating of the dialectical processes described in the PCS framework I have applied.

Being able to reciprocate by helping to blur the boundaries between the assumed competence and value of youth, and the assumed incompetence and worthlessness of dependent old age, had been in the gift of participants in both countries, in the form of the passing on of wisdom and skills to future generations. For example:

- Kannamma, Subbulakshmi and Doris had demonstrated expertise in traditional crafts;
- Vadivu had the experience and knowledge necessary to lead prayer rituals and help people understand their significance;
- Lakshmi was competent in a number of languages and had an understanding of political history that spanned almost all of the twentieth century; and
- Ken remained competent, willing and enthusiastic to instruct in engineering, joinery and so on.

But evident in both sets of responses was also a lack of any formally created initiatives for realising that potential in ways which could have contributed to facilitating their affirmation as 'good citizens', or even citizens at all, according to ageist discourses that deny older people that status, (Midwinter 1990). Where such intergenerational reciprocity could be seen to operate (at residential home A in Chennai), it had not been at the instigation of the staff, but at that of the residents themselves. Given that both the older residents and the young female students supported by the owner's benevolence were spending time at the same premises on a regular basis, it seems that this may have been the result of a serendipitous matching of need and resource, rather than any real commitment to facilitating reciprocity as a foundation for spiritual well-being. Yet six of the seven Indian elders, and five of the seven UK elders, indicated that they would have welcomed opportunities to have been involved in teaching or helping schoolchildren, students or young families (see Table 5.1)—a finding which:

(a) provides a challenge to dominant bio-medical and ageist ideologies that presume older people to have nothing useful, and no desire, to contribute in that way; and
(b) endorses phenomenologically grounded inquiry as a vehicle for hearing perspectives which have the potential to challenge dominant discourses, and which might otherwise not have been heard, or given credence.

I would not wish to pathologise care providers, however. Rather, I suggest that the perceived lack of attention to connectivity, reciprocity and spirituality reflects and reinforces the body of literature that highlights the power of bio-medical and ageist discourses, in combination, to mutually reinforce the message that looking after the body and its physical functions is what care is assumed to be about (Twigg 2004; Gilleard and Higgs 2011). While dependent older people such as Kannamma, Lakshmi, Megan and Ken may have perceived themselves as having been constrained by the negative stereotyping of frail older people as useless, it was also in their power to have 'reconstructed their care biographies' (Cook 2008). For example, they could have challenged that negative stereotyping by lobbying for intergenerational initiatives to be widely and formally incorporated into eldercare institutions, on the basis that dependency need not preclude reciprocity.

In conclusion, and with reference to the ways in which social space is significant for understanding the impeding or promoting of reciprocity, it is clear from the findings that this had been compromised to the extent that the participants' relationships in social space would have been limited to being predominantly with care staff or fellow residents. As a consequence, intersubjective feedback is likely to have been informed predominantly by visible and assumed deficit, rather than on aspects of competence, or the potential for spiritual and intellectual growth. A broader range of feedback, such as that from people who had known the participants before they had become dependent on formal care, would have had the potential to challenge this negative association by helping participants to maintain a positive autobiography. However, the necessary facilitating of such connectedness was not evident in the findings. And without that connectedness:

(a) the 'fusion of horizons' needed for learning and personal growth, (Gadamer 2004), is unlikely to be facilitated through relationships of reciprocity;

(b) the potential for the association between old age and stagnation to be challenged is undermined; and
(c) the spiritual well-being of those older people dependent on formal care is likely to continue to be compromised.

Section Two

In the previous section, I discussed how recognising the significance of 'being-with-others' (Moriarty and Butt 2004; McMunn et al. 2009; Attig 2011) for the self-esteem and spiritual well-being of dependent older people offers a challenge to atomistic approaches that do not take the context of their experience into account. In this section, I address the significance of the temporal issues that emerged from the findings, not only relating to change at a personal level (ageing, relocation and so on), but also to wider changes, such as those at the level of ideas, policy and power relationships.

In addressing the second of my research questions:

> In what ways is the concept of social time significant in accounting for how reciprocity in the care of dependent older people in receipt of formal care is promoted or impeded?

the following themes emerged as significant. As with the first research question, they incorporate those themes that arose directly from the data analysis process. If reciprocity is not to be impeded, then there needs to be recognition that:

- older people have aspirations for the future, as well as experience of the present and memories of the past (Baars 2007);
- there is continuity and change in terms of our values as we age; and
- age cohort is significant, in that prevailing normative assumptions about ageing change over time and between generations (Bytheway 2011).

And so to the significance of the first of these themes.

Neglecting Aspiration

Much of what was said by the participants focused on the relationship between their current situations and their past—how they experienced a mismatch between how they had felt about themselves before and since becoming ill or frail. Less was said about the interplay between present and future—that is, on how their current situation might be affecting how they saw themselves in the future (Sartre 1963).

Aspiration was not absent altogether—it did feature in Megan's story, for example, in terms of her aspiration to age with her mental faculties and values intact informing her present focus on keeping her brain active by doing crosswords and playing solitaire as a daily routine. Similarly, Muthu's hope to ultimately go to 'paradise' (the place where he assumed he would be rewarded for reciprocity) could be seen to inform his current desire to prove himself in that respect by engaging in the acts of reciprocity that he described as 'social work'. While Kannamma and Ken implied aspiration when they referred briefly to unrealistic hopes to teach, and be involved in the affairs of the social club, only Megan appeared to be acting on them, taking positive steps to carve out a role for herself which would allow for personal growth and the harnessing of her skills and experience for the benefit of others. I wondered whether her outlook, unique as it was within the whole sample of 14 participants, could be attributed to her not being influenced by the overprotective cultures seen to be operating in the various residential care facilities I visited. However, neither Ken, nor Harold, had been living in those environments either, yet they appeared to be acquiescing to the same assumptions reflected by those in residential care—that is, their 'useful' days were largely over. What may have been more significant, therefore, than the residential/domiciliary care dichotomy was, perhaps, the presence or otherwise in their daily lives of health or social care workers, whose primary (or sole) focus is likely to have been to address physical needs, and protect from harm, in the here and now (Thompson and Thompson 2001). Although Megan had been assessed as dependent enough to warrant support services, she had declined them, choosing to live what was considered by others as a 'risky' lifestyle, and making conscious efforts to live alone by her own rules, for as long as she could. It is possible, then, that it was:

(a) her refusal to be defined by her age (88) or her health problems (significant and potentially debilitating); and
(b) not being reminded that she was ill by having carers visit her on a regular basis,

that accounted for the difference between her perception that personal growth was still attainable, and the perception of the other participants that personal growth was desirable but no longer possible. In considering why Megan alone appeared not to have internalised the assumptions embedded in the medical model that vulnerable people need protecting rather than empowering (Thompson and Thompson 2001; Bond and Cabrero 2007), it may have been her character, life experiences and personal values that accounted, at least in part, for this anomalous finding.

For the most part, though, the narratives did not reflect future aspiration. Several, including Venkatram, Vadivu and Edna had talked about a general sense of contentment with the life they were currently living—a willingness to live in the moment—and Doris appeared to write her future off altogether;

> There's not a lot in the future actually. I can't walk very well but unless my daughter and her husband come to take me out in the car, I don't go out ... but, I mean, I'm happy enough here except there isn't a lot to do.

But if those responses are considered in the light of the insights provided by Thompson (1998), who has based his discussion of the progressive-regressive method on Sartre's work, and to 'thrownness' as described earlier Heidegger (1962), then this lack of awareness of (and reluctance to take account of) a future dimension can be seen as problematic because it has consequences for how self-esteem is maintained in such circumstances. Furthermore, it also has implications for how dependent older people's self-concept is maintained. That is, their sense of who they are and want to be—an important part of their spiritual dimension. If we accept Heidegger's and Sartre's premise that our sense of who we are in the present is constructed partly by our experiences in the past, (who we have considered ourselves to be), but also by projections we make into the future (who we aspire to be) then, if the future dimension is not recognised, the potential for growth is unlikely to be recognised either, and the potential for people to feel estranged from their ideal or preferred self heightened.

The implication of this for an exploration of reciprocity lies in the fact that, if dependent older people and those who support them focus only on the present (and, by association, on their weaknesses to the exclusion of their strengths), then the senses in which they are perceived as 'givers' may become lost or minimised. As a consequence, defined by their dependency, they become seen as predominantly 'takers'—and, as Harold described himself, 'past it'. Where the future dimension is absent from the dynamic, there exists the potential for the present to be perceived as the endpoint of a continuum of personal growth and assumed value. This appears to leave only one option—to accept that one's identity as someone who can reciprocate has been compromised—and to deal with that as a loss issue. However, as Doka (2001) has argued, many forms of loss are disenfranchised—that is, they are not recognised and validated as loss experiences. And so, though older people themselves may perceive the challenge to their identity as valued people as a form of loss, the experience may not be validated as such by others who assume incapacity in old age to be normative. But where a future dimension *is* recognised by the providers or facilitators of care, it reflects an understanding that the older person has not arrived at an end point, but remains on a journey of personal development until they draw their last breath. And where *that* is recognised, it allows for:

- an appreciation of how identity as a 'giver' can be maintained, rather than lost, in the face of the challenges to it;
- new ways, and forms, of reciprocity to be promoted and facilitated; and
- the significance of the changes for people's sense of identity and self-esteem to be explored through processes of meaning-reconstruction and reframing (Neimeyer and Sands 2011).

Ken's responses were interesting in this respect in that, though he was dealing with the fact that he was terminally ill, and mourned the loss of opportunities to feel useful, he also talked about how his helping others (such as the new handyman at the social club, and the grandson who had joined the caring profession after being Ken's carer) had contributed to his current self-esteem, and had highlighted to him how it was important to have an ongoing sense of 'usefulness' if he were to continue to feel good about himself. However, while it seemed that Ken had both the desire and skills to continue reciprocating, these were not recognised by those providing respite care at the facility he attended:

> it's a lovely place but I didn't like it because there was a lot of old people—well I'm old—a lot of old people, and all they did was falling asleep and all this. There was no activities.

As such, therefore, his aspiration to continue using his skills to help others did not appear to have been facilitated in a meaningful way—in that the only opportunity on offer had been involvement in craftwork, which appeared to be the making of trinkets and ornaments for the sake of being occupied, rather than being linked to his particular interest or skills:

> you know, they're giving you something to do.

Given his history as a skilled craftsman and his aspirations to continue passing on his skills and expertise to others, it seems to me that it would have been easy enough for an element of this to have been incorporated into such activity sessions—Ken co-leading a session (even one in a subject of his own choosing, perhaps), or at least helping others where he had the experience and dexterity to help his less experienced or confident peers, Ken may then have been able to reframe the experience as one of being *valued*, rather than just being *busy*. That is, one of reciprocity, rather than merely engagement.

That his narrative reflects that he was not enabled to do so, supports the assertion by (Knight and Mellor 2007) that activity sessions often neglect unique needs and strengths. It seems to me that his care, both at home and in the care facilities he attended, was informed by bio-medical and ageist discourses that led to a focus on keeping Ken physically comfortable and busy in the present, but neglected to take account of his having a future, in the sense that he wanted to be enabled to continue doing what had contributed to his self-esteem in the past, so that he could 'live while he was dying' (Saunders 2002)—that is, living in the sense of growing and continuing to be appreciated for his contribution to society, rather than just existing. It was clear from his responses that there were still things he would like to have been doing that held more significance for him than, as he described it, 'painting by numbers' and 'making flowerpot men'. But for those aspirations to have been facilitated, someone would have needed to have been aware of them and, as his responses also indicated, no-one had thought it relevant to ask.

Again, it is not my intention to apportion blame by pathologising individuals as uncaring. Rather I use this example to highlight once again how not recognising and challenging a culturally embedded assumption (that it is acceptable, on the grounds of age alone, to neglect such an important aspect of personhood as aspiration) can reinforce the legitimacy of age as a basis for social division (the S level of PCS analysis).

Their narratives suggest that many of the participants were also accepting of this limited perspective on their needs and, as a result, were 'living down' to the stereotype of old age, rather than challenging it by pushing for their aspirations to be facilitated. And for most of them, those aspirations could have been easily facilitated, which would then have had the potential to challenge commonly held negative assumptions about the 'value' of older people dependent on formal care. For example, Subbulakshmi had aspired to play a role in the domestic upkeep of the residential home. Despite her failing health, this aspiration could have been addressed by inviting her to act in an advisory capacity, but was not, and to her spiritual detriment because she found it distressing to be conceptualised as a burden.

Similarly, where Eddie had felt keen and able to support ex-servicemen, but impaired mobility had made leaving his surroundings difficult, this could have been facilitated through the medium of virtual communication, but was not, and to his spiritual detriment, because he considered that others who might have benefited from his contribution were being denied his help. It seems that their 'place-in-the-world' had become an ascribed one, based on assumptions about who they *should* be as older people, rather than a reflection of who they considered themselves to be, wanted to continue being, or aspired to become.

There is also a temporal element to the acquiescence that was reflected in the majority of the narratives. For some, this appeared to be related to a perception that matters had gone beyond their control—several expressing having found themselves in situations they had never expected to be in, as is implicit in the concept of 'thrownness' (Heidegger, ibid). For example, Doris, who had lived alone in another part of the United Kingdom, suddenly 'found herself' several hundred miles away, in an institution, and disconnected from the community in which she had been valued—apparently with no obvious opportunities for establishing herself as competent and 'useful' in her new situation, where her identity as frail had already been framed. Similarly, Kannamma, who had hoped to live and teach with her son in America, instead found herself living in an institution where her aspirations could not be realised, and where the significant challenge this would have posed to her spiritual well-being appears to have gone unrecognised. Her reframing of herself as a valued person through the giving of gifts as a benevolent act may therefore have been an attempt to:

(a) make sense of a significant, though disenfranchised, loss experience (Doka 2001) and
(b) reconcile the 'estrangement from self' (Fromm 1955), which may have been a consequence of the impeding of opportunities to function as the person she perceived herself to have been in the past, and wanted to continue being.

The findings have supported the assertion by Komter (2007) that the meaning associated with a gift is important, by highlighting that the significance of reciprocity for the spiritual health of older people dependent on formal care may lie in their being *able* to reciprocate, rather than in *how* they reciprocate. Furthermore, though the act itself may have seemed insignificant from the perspective of others, it may have held significance for her in terms of contributing to the sense of 'mastery' (Chokkanathan 2009) espoused at the C level of shared understandings within Indian culture.

As I discuss in more detail later, some of the differences noted in connection with aspiration to continue in valued roles can be seen to relate to different discourses (pertaining to religion, for example) operating in the two contexts. For example, several of the Indian participants referred to the future being out of their hands, and related their aspirations only to being able to live out what they believed their god had in store for their future. This was reflected in Vadivu's comment:

> I have no children, my husband was already died so there's no-one depending on me so I don't owe to anybody—so I came here. What's left is belonging to God.

But while resignation to a life not of their own choosing was featured in both sets of responses, this did not appear to be attributed to fate or a deity by anyone in the UK sample, where any deference, such as that shown by Doris, Maureen and Harold, was to the decision making of health, social work or social care professionals. That the concept of aspiration was absent from formal procedures of assessment of need, (the experience of all but one of the UK participants) is particularly disconcerting because it suggests that the neglect of spiritual and existential needs is more widespread than can be explained by a lack of individual awareness. This is addressed further in Chap. 8.

Personal Values: Continuity and Change

Although ageist ideologies would have us believe otherwise (Butler 1987; Nelson 2002; Thompson 2005) the lives of older people dependent on formal care remain dynamic and open to change (Biggs 1999; Hill 2005; Gilleard and Higgs 2005). We all change in some respects as we age, and stay the same in others, and this is true of our values (Moss 2007). As our values influence how we behave towards others, they can be said to be part of our habitus (Bourdieu 1984). As already discussed in Chap. 2, habitus, by its definition, helps define uniqueness—it is what makes a person that *particular* person and not a stereotype. In terms of reciprocity, most of the participants expressed being able to give back in some way as contributing to making them the person they perceived themselves to be—that is, it was an aspect of their habitus. This is evident in Megan's describing of her support to others as:

> that's me!

and Edna's reference to cheering people up as:

> (I'm) that sort of person, although I say it myself ... I love helping people.

Muthu's use of the expression:

> a helping tendency.

to describe himself also seems to be in a similar vein. However, the potential for this to be challenged by ideas about how they *should* behave, can be said to be linked in part to normative assumptions about age identity (Biggs 1999).

The relationship between actual and perceived age identity is reviewed by Park et al. (2010) in their chapter of the British Social Attitudes Survey 2010–2011, where they make the point that: 'age is one of the most important bases on which respondents will express feelings of commonality with others in their selected identity group' p. 195. They distinguish between 'passive' social identities (where there is recognition of commonality, and apparent acceptance of the identity one has been assigned by circumstances) and 'active' social identities (which appear to incorporate the capacity to reject externally assigned characteristics). Eddie's and Megan's resistance to the label 'old' (because they did not see themselves as acting in ways they associate with their peers) would appear to be well explained by the concept of an active social identity. Furthermore, Megan's incorporating of new ways of reciprocating into her self-identity as a valued person would appear to support the premise that habitus is generative (Garrett 2010)—highlighting that the precluding of reciprocity in dependent old age is not inevitable, but open to change if the will exists. As with empowerment-based feminist discourses which argue that women do not have to be the victims of patriarchy (Lister 2003; Hague and Mullender 2006), both Megan and Eddie's narratives reminds us that older people do not have to be the passive victims of ageist ideology and, therefore, can be seen to support the work of Grenier and Hanley (2007) on resistance and frail identities and that of Cook (2008) in relation to the reconstructing of positive identities when older people move to care homes.

For the most part, though, the participants in my study appeared to be adopting a 'passive' social identity in terms of age, by accepting the physical and ideological limitations placed on their aspirations, even where this compromised values relating to the desire to be conceptualised as valued citizens, neighbours or friends. And by doing so, one of the dialectical processes described in the PCS framework can be seen to have been operating—in that the pathologising of *all* older people (through blanket assumptions at the C level that neglect uniqueness and overlook strengths) was informing their perception that they should receive, but not give back—thereby reinforcing their subordinate position at the S level.

The point made in Sect. 1 about perceiving oneself to be a different person than that before becoming dependent also has relevance here, where the focus is more specifically on temporal issues. The findings have indicated a difference between the perceptions of the UK and Indian elders in terms of the relationship between former and current identities, in that reciprocity as an *enduring* espoused value was not as evident in the narratives of the UK elders as it was in those of the Indian elders. Rather, a more typical theme was that of a rift between former and current identities. So, for example:

- Harold, having been a respected teacher, now considered himself to be of little use and had low self-esteem—yet several of the Indian elders continued to teach;
- Eddie, having been a health practitioner, was unable to utilise those skills—yet several of the Indian elders continued to do so; and

- Doris, having been a competent homemaker and secretary, found herself redundant and bored—yet the Indian elders continued to take responsibility for keeping their immediate environments clean and tidy, within the limits of their abilities.

It may be significant in terms of their spiritual well-being that, across both samples, the participants expressed their dependency most often in terms of illness, rather than as needing help. When they described themselves as ill, the perceived need to reciprocate may have seemed to be less important to them than it had been. Doris, for example, justified her admission into residential care by reference to her heart attacks. She had described how much self-esteem she had gained in the past from what she could do for family and work colleagues but, since moving into residential care, appeared to resent others' reliance on her, only engaging superficially with other people in order to stave off boredom, and not appearing to feel any compunction to be helpful to others for *their* benefit.

Kannamma, too, had expressed expectations of being looked after without an obligation to give in any way to fellow residents or staff. It may have been significant that both Doris and Kannamma had been paying for their care, and, as such, the findings would support the premise inherent in the commodification of care thesis, (Ungerson 2000; Garey et al. 2002), that paying for care has implications for the understanding of carer/cared-for dynamics. In terms of the implications of this for an appreciation of the spiritual well-being of older people dependent on formal care, it may be particularly significant that the narratives of both Kannamma and Doris appeared to reflect a change in their personal values. For example, while Doris's narrative of her earlier life had reflected how she had valued reciprocity (to the point of citing her parents as role models in this respect), that of her life after becoming dependent reflected a much more self-interested individual. I have no direct evidence, but, in light of research relating to mastery (Pearlin and Schooler 1978; Beel-Bates et al. 2007; Chokkanathan 2009) surmise that this may have been an attempt to regain a sense of control in her life—a way of re-asserting her independence by declaring that she would live life on her own terms, and in her own interests, after a lifetime of looking after other people's interests.

The Significance of Age Cohort

A further aspect of the significance of social time for the impeding or promoting of reciprocity is the extent to which understanding, and expectation, of old age have changed during older people's lives.

Given that reaching 100 years of age, in the UK at least, is no longer unusual (a statistical bulletin from the Office for National Statistics estimating it in 2008 as 11,000 and rising), and that the beginning of old age is typically defined in line with eligibility for old-age pensions, the period defined as 'old age' can therefore often span 30 or 40 years—thereby incorporating at least a generation of difference in social attitudes, including those relating to expectations of reciprocity in old age. Thus the life experience of someone *currently* defined as being in advanced old age is likely to be different from that which someone at the beginning of the old age spectrum now will experience in *their* advanced old age. Demographic and social policy changes in India, have raised the possibility that attitudes to ageing, and expectations of older people, may be changing there too (Datta 2008).

Returning to the UK situation, while it may be true that those entering old age currently are doing so in an era characterised by consumerism (Gilleard and Higgs 2005), those whom I interviewed would have spent the earlier stage of *their* old age (when they were less dependent) in different political, social and financial climates from the current age cohort of independent older people. And, with changing age cohorts, come changing social rules, norms and expectations, so that assumptions cannot be taken for granted (Timonen 2008). For example, Doris's disappointment at the lack of manners shown by younger people demonstrates an example of how behaviour can be said to hold different significance for different generations. Doris may, for example, have had different expectations of younger people, and perhaps been less offended by their behaviour, if she were experiencing her 'young' old age in the current age cohort of independent older people who are experiencing old age in a social climate characterised more by individual responsibility and self-interest than her own earlier life had been (Beck and Beck-Gernsheim 2001).

This has implications for expectations around citizenship duties and obligations of reciprocity, as it could be argued that, with changing cohorts, may come changing expectations of what it means to be a 'good' citizen, and for whether reciprocity is incorporated into prevailing norms about socially valued behaviour (as can be seen in the UK in the call to communitarianism that underpinned New Labour's 'Third Way', and the current government's vision of 'The Big Society'). It was apparent from several of the Hindu elders I interviewed, that the concept of good citizenship continued to have currency in their social situation, but contained contradictions. On the one hand, their narratives reflected a duty to help others where the capacity existed but, on the other, also an expectation that one should take responsibility for one's own actions and not seek help from others unless necessary (Hodge 2004). This may have informed Venkatraman's comment that:

it is said that to not trouble others with our worries will bring happiness …

Though current conceptualisations of old age (informed, as I have suggested previously, by debates about positive and active ageing) can be seen to incorporate a challenge to the assumption that dependent old age and reciprocity are incompatible, a significant proportion of the participants in this study had experienced *their* old age against a backdrop of opinion, informed by bio-medical and ageist discourses, which assumed that they are.

Renzenbrink (2010) speaks poignantly of how issues of age cohort can be implicated in the impeding of reciprocity by there being a lack of access to the modern communication technologies which offer the potential for connectedness. She has argued that, in her experience of eldercare in Australia, the need for frail and dependent older people to connect with others is rarely recognised or facilitated in care homes.

I am not aware of any research that quantifies the extent to which connectedness between isolated older people and wider communities is routinely facilitated via communication technologies in either the UK or India, but I suspect from my own experience, and the inferences of social work colleagues in Chennai, that it is often neglected. I was therefore not surprised that the findings reflected an almost total absence of reference to such communication channels, other than Megan's mention of her capacity to send emails. The care facilities I visited in India were being run on shoestring budgets, and so this was understandable in that context. However, given the relatively low cost involved, that it was not reflected in the UK narratives raises the question of whether this had less to do with the cost *per se*, than the power of ageist ideology to portray dependent older people as not worthy of investment because of the erroneous basis that they have nothing to give in return.

Section Three

While the first two research questions have implicated meaning making, the third:

> In what ways is the concept of meaning making, as it relates to individuals and their spirituality, significant in accounting for how reciprocity in the care of dependent older people in receipt of formal care is promoted or impeded?

addresses it more overtly by exploring the links between reciprocity and the sense of well-being associated with ontological security—that is, knowing one's place in the world, feeling comfortable with it and valued because of it. Of the many aspects of meaning making that arose from this research I have chosen to build the discussion around four issues that I feel are particularly significant in terms of the potential to impede reciprocity:

- a focus, in assessment and care delivery, on the physical dimension at the expense of the spiritual;
- a lack of recognition that health crises can also constitute existential crises – that is, a challenge to self-perception as a competent person;
- the perception of deference as reciprocity; and
- the existence of a shared 'world-view' that espouses it as a virtue.

Neglect of the Spiritual

As already alluded to, one's spirituality relates closely to one's sense of self—who we think we are, or would like to become (Goldsmith 2004; Moss 2005), but also how we fit into the wider world. When we are engaging in soul searching it is often an internal dialogue but, as social beings, it is almost impossible to have that dialogue without being affected by external influences. For dependent older people doing that soul searching, prevailing cultural messages about their lack of competence and value are not in short supply (Jewell 2004). And so, taking on board the premise that we need positive affirmation in order to flourish (Moss 2008; Attig 2011), it makes it all the more important that those negative assumptions are challenged if spiritual well-being is not to be compromised. The findings from this study suggest that spiritual flourishing, if considered at all in care planning or delivery, was very much secondary to physical flourishing, which supports the premise that this dimension is often lacking (Mackinlay 2001; Moss 2005; Tanner 2010).

For example, reference has already been made to the aspirations of at least half of the participants to engage *meaningfully* in the communities in which they were living. Yet, for a variety of reasons, little attention appears to have been directed to making these aspirations attainable.

In Subbulakshmi's and Eddie's case, they perceived this as having a direct correlation with the risk management policies of the institutions in which they lived, but the others did not mention being 'prevented' as such. The narratives of Ken, Harold, Maureen and Muthu spoke to me more of a lack of *recognition* by care facilitators and providers that having a valued role mattered to these people, than of a set of risks to be managed. So, in response to the question 'did they get that affirmation as valued people', I would have to conclude from my findings that, apart from Megan's relationship with the voluntary agency she had been called to work with, they did not—at least not from their care providers or facilitators. Had the implications for their self-esteem been recognised then I would have expected them to have referred to at least some attempt to help them 'stay me', but none was in evidence.

I also referred in the literature review to the potential for a process of welfarisation to portray older people, especially significantly dependent older people, as a problem to be addressed, rather than as individuals experiencing problems (Fennell et al. 1988) and the findings can be seen to reflect that process of dehumanisation. Though actual criticism was expressed only by Ken, implicit in the findings was the perception by the participants that there was an absence in their care provision of the 'care-as-relating' relationships described by Bowers et al. (2001)—defined by the presence of reciprocity between caregivers and care receivers, and indicating respect for how each party can contribute to the well-being of the other. More typically represented, in that concerns relating to physical risk appeared to override other well-being issues, were relationships akin to those described earlier as 'care-as-service' (Bowers et al. 2001). To the extent that the care focus was instrumental, they reflected the 'I-It' relationships described by Buber (1958) because they evidenced a lack of respect for personhood.

Where such relationships exist unchallenged they can be seen to constitute an impediment to reciprocity because, where the significance of reciprocity for well-being is not recognised, then a responsibility to facilitate it is unlikely to be recognised either. And, with reference to PCS analysis (ibid), where processes of welfarisation, dehumanisation and alienation (in the existentialist sense) inform understanding of the spiritual dimension of dependent older people's lives (the C level), they have the potential to both a) reinforce the legitimacy of marginalising older people generally (at the S level) and b) inform attitudes and behaviour at a practitioner level (the P level). As I go on to suggest in Chap. 8, eldercare practitioners are well placed to promote the recognising and addressing of the spiritual need to feel valued, but this positive potential was not reflected in the findings of this study.

Health Crises as Existential Crises

Health crises are far from unusual in old age (Bond and Cabreo 2007) especially when associated with poverty (Coote 2009), and typified the life experiences of those I interviewed, often having been a precipitating factor in their admission to residential care. For some, the crisis had been sudden. Lakshmi and Vadivu, for example, had been admitted because they had fallen and broken bones, and could not manage to look after themselves while incapacitated by their injuries. Kannamma's admission had also been as a result of a medical emergency, which had seriously impaired her mobility. For the remainder, the decline in their health had been associated with increasing frailty and chronic conditions such as Muthu's and Subbulakshmi's breathing difficulties, Ken's kidney failure, Eddie's, Kumari's and Doris's heart problems and Maureen's depression. For the majority, then, their increasing dependency on others had been a slower process. However, regardless of whether or not they had constituted an emergency, these admissions may nevertheless have still constituted a crisis point by undermining previously held 'certainties' in their lives and their confidence in their coping skills (Thompson 2011b).

In terms of impeding or promoting reciprocity, it is significant that their health crisis (in the sense that they could not continue as before, and therefore required care support of some type), in causing an interruption to their independent living, also had the potential to contribute to an existential crisis—that is, to lead them to question what it meant to be 'me'. Most appeared to have experienced an existential crisis that had affected their sense of self as someone who could continue to reciprocate although, once again, Megan's narrative was atypical of the findings. For example, she had not seen her declining health as something that necessitated a change in the ways, or extent to which, she planned to continue giving back to her community. While her frequent falls may have constituted health emergencies on some occasions, she did not appear to perceive of them as crises because she did not allow them to challenge her self-image as someone still competent to reciprocate, despite the risks to herself.

Others, however, did appear to see their health problems as crises, in the sense that their biographical continuity had been disrupted, and their spiritual well-being adversely affected by having the 'certainties' in their lives put under threat. The narratives of Vadivu and Maureen in particular reflected the adoption of a passivity associated with a 'sick' role (Daniel 1999). In doing so, they may have been trying to resolve a troubling internal contradiction by classifying themselves as ill, rather than dependent, so that they had a different set of normative rules to inform the reconstruction of their self-image as people who were not *expected* to contribute or reciprocate. That is, conceptualising themselves as sick may have helped to make being 'done to', and the loss of an active role easier to accept. It may have seemed less of an 'assault' on their spiritual identity as older people *able* to reciprocate by engaging in socially validated roles, but having opportunities to do so neglected or impeded by others.

Managing contradictions between their own perceptions of self-worth, and those of others, appeared also to have been an issue for Eddie and Harold who, though recognising that they did indeed have serious health problems, were distressed by assumptions that they would drop their aspirations to continue being of use to others, and adopt the more passive role associated with sickness. In Harold's case, his carers may have felt that they were acting 'appropriately' by relieving him of household chores but, from Harold's perspective, they were challenging his sense of himself as competent, which appeared to cause him distress. As he told me:

> I don't mind cooking at all. I'd rather cook for myself than have anyone to cook for me. They can do [take over], sometimes—you know, suffocate you, if you let things get that far ...

This tension appeared to be even more distressing for Ken who, having accepted that he had very serious health problems that compromised his ideal lifestyle, had engaged in a re-evaluation of who he felt he was, or had become. Although accepting of the need for formal care support, it appeared to trouble him that what he perceived as their disregard for issues he felt strongly about, (the wastefulness of washing his pyjamas and bedding every day, for example), appeared to trouble him because *his* perspective on matters relating to *his* household was not being sought—and what he could give to the situation, (opinion based on years of life experience) ignored.

I return, in Chap. 7, to the epistemological issue that this raises in terms of the lesser value accorded his version of 'the truth' of his situation.

In addition to the existential challenges that health crises may have constituted, the narratives had also reflected practical issues that had the potential to impede reciprocity by limiting opportunities for the acquisition of social capital which, as has already been established, has been positively linked with well-being (Bourdieu 1986; Putnam 2000). Maureen, for example, had expressed that she had few opportunities to feel socially validated, such that she experienced low self-esteem:

> not being worthwhile as a person, really.

Arguably, this could have been supplanted by the self-esteem she might have gained from being able to play a useful role in the community surrounding the residential home, had someone been made available to support her to find a situation where she felt her skills could be employed, and to help her build up her confidence to do so. Evidence in the form of Maureen's aspiration to engage with others in her community (an aspiration echoed by others, including Muthu and Eddie) supports that literature which associates social engagement (Gardner 2010) and connectedness (Attig 2011) with well-being—particularly, I would suggest, spiritual well-being (Moss 2005; Fernando 2007).

The process of latent thematic analysis had identified 'feeling valued' and 'having opportunities to reciprocate' as being *particularly* significant to the participants. That opportunities had been few for both sets of participants may reflect the consequences of the medicalisation of old age (Powell and Biggs 2000; Westerhof and Tulle 2007) such that the findings support existing theses that highlight the associating of old age with illness and incapacity. However, it seems, that, for the most part, the Indian participants may have been less affected than the UK elders by assumptions that declining health would necessarily affect their capacity to continue helping others.

For example, though Venkatraman had made a conscious decision to withdraw to a life of personal spiritual enlightenment, he did not appear to have any doubts about being able to help others if he were to so choose. Muthu, also felt that he would be able to continue helping others if the opportunity were to arise. The remaining Indian participants all described ways in which they had changed the form, or manner of their acts of 'giving back', in the face of the existential challenge to their sense of being a 'good person' that their health problems had faced them with. This fits with what appears to be expected of them as Indian elders—in the sense that, as reciprocity is considered to be a life-long commitment related to 'service to humanity' (George 2008), then should it become impossible to enact it in one way for any reason, another would be sought so that spiritual well-being would not become compromised.

By contrast, the narratives of the UK elders reflected more of a sense of acceptance that incapacity and declining 'usefulness' were becoming a feature of their lives, despite, in many cases, their being unhappy about it. Part of the reason for these differences may lie in the power of well-defined and shared world-views to influence values and aspirations, and this is discussed in more detail later in this section.

Reciprocity as Deference: A Case of Self-Betrayal?

For a number of the participants, reframing their acquiescence as, in itself, an act of giving, appeared to provide a way for them to make sense of the changes they were facing. Though this did constitute an act of reciprocity, and would have preserved a degree of dignity, I would suggest that it would be considered a 'degraded' form of reciprocity. That is, though the disempowerment it incorporated was *self*-disempowerment (the P) level, I would suggest that it was, to an extent, forced on them by unequal power relations (the S level.) As such, this element of their narratives supports existing literature that highlights deference as a form of reciprocity (Beel-Bates et al. 2007). While it may have appeared from an outsider perspective to have been to their own cost in terms of preferred lifestyle, their own narratives reflected that decisions made in other people's interests had contributed to their own peace of mind and, therefore, spiritual well-being. For example, Doris reported that she had not really wanted to leave her own home, and wished that things had been otherwise. Similarly, while she had been reluctant to enter a care facility after more than nine decades of independent living, Lakshmi had apparently done so at the request of her benefactor and friend—an act which also could be construed as an act of reciprocity in the form of the gift of peace of mind.

And while Doris and Lakshmi's reciprocity in the form of deference was implied, rather than clearly articulated, Muthu's wish not to be a burden was more overtly expressed:

I'm not a selfish person... Helping tendency is not to be a burden. I want to be active.

In analysing the findings, I found myself asking the question: 'why would these people sublimate their wishes to the concerns of others, when latent thematic analysis had highlighted that, for almost all of them, reciprocity had been identified as a 'feel-good' factor in their lives before they had become dependent on others?' As already discussed, this may be addressed in part by the existence, and internalisation, of discourses relating to:

- the moral legitimacy of using age as a basis for social division, and a justification for treating older people less favourably than others (Powell and Biggs 2000; Thompson 2005; Neuberger 2009; Bytheway 2011);
- blanket assumptions that dependent old age precludes reciprocity (Antonucci and Jackson 1989; Boerner and Rheinhardt 2003; Breheny and Stevens 2009); and
- professional power and decision making (Beresford and Croft 2001; Powell 2006).

It is possible that, in light of the power of such discourses to suggest that deferring to the interests of others is 'how it should be', those participants who did so may have been employing it as a strategy to preserve their dignity for fear that those decisions might have been made anyway by others who deemed them to be 'at risk'. The kudos associated with professional status (Hugman 1991; Thompson 2007) may have contributed towards making those providing support appear more powerful than they were to the vulnerable, but competent, older people in this study who, in reality, retained the right in law to make their own decisions about risk taking (Wiseman 2011) but may have perceived it to be otherwise.

A Shared World-View?

It had become clear during the interview process in Chennai, that a social expectation existed in that place, and at that point in time, relating to the expectation that one should give financial assistance to others in need, if one had the means to do so. This expectation of benevolence was evident from the numerous and regular appeals published in the local and national newspapers for help to finance surgical procedures, the purchase of expensive medications or to respond to unforeseen emergencies that people could not afford to address. The perception of reciprocity as beneficial to spiritual well-being may therefore have had an extra dimension for the Indian participants that would have been unlikely to have been in the experience, and therefore the consciousness, of the UK participants.

Again, this demonstrates how cultural meanings and structural arrangements (the C and S levels of PCS analysis) can be seen to be linked in a dialectical relationship. That is, Hindu communities in India—where age currently does not appear to be particularly significant as a basis for social division—the moral duty to reciprocate that is associated with Hinduism (Mehta 1997), may more readily be expected of *all* adults—including dependent older ones. In contrast, where age *does* constitute a significant basis for social division—as is the case in the UK—then expectations of younger citizens may not necessarily be seen as relevant to dependent older people.

When considering the significance of what the UK elders shared with me, it was more difficult to detect a shared way of looking at the world, or a shared set of clearly espoused values relating to reciprocity. I had not asked about religious affiliation, and only one made theirs known to me during the interview. What appeared to be more significant to the UK elders in terms of meaning making was whether reciprocity had been espoused as a core value within family culture—as a moral norm that set expectations of reciprocity (Gouldner 1960). For example, Doris had held her parents up as role models—describing how they had shared what little they had during wartime rationing—reflecting that the value of giving and sharing had been promoted positively within her family culture. Similarly, 'give and take' had been a philosophy that had informed married life for both Harold and his wife, and for Megan and her husband, as in both households, paid work and household duties had been shared. It seems that reciprocity had therefore been an espoused family value for them too.

Section Four

> In what ways is the concept of meaning making, as it relates to discourse and institutionalised patterns of power, significant in accounting for how reciprocity in the care of dependent older people in receipt of formal care is promoted or impeded?

My fourth and final research question invites consideration of the broader cultural and structural contexts which have an impact on:

(a) how individuals strive to make sense of what is happening to them when their sense of self as competent people with the capacity to give, as well as take, is called into question; and
(b) whether others see reciprocity as something that should be on the eldercare agenda.

I argue that, where studies of reciprocity focus exclusively on relations and processes operating at a psychological level, they neglect to take account of the complex and changing sociological dimension of human life and relationships. In adopting a psychosocial perspective, this study takes into account that:

(1) Meaning making does not happen in a vacuum—both those who provide care and those who receive it are subject to a variety of accounts of how life is and should be.
(2) Opportunities for, and expectations of, reciprocity will be mediated not only by assumptions about old age, but also its articulation (operating at the S level) with other socially significant social divisions, including ethnicity and gender (Arber et al. 2003b; Boneham 2002; Jacobs 2010).

I had doubts whether an understanding of these broader contexts would readily emerge from the interview data, as both the methodology and the phenomenological premise of the study, focusing as they do on biography and perception, lend themselves more to an exploration of the dialectic between personal experience and shared meanings, than to the structural issues that also have an impact on lived experience. Furthermore, I would not have expected the participants to have necessarily been aware of these issues, or to have articulated them in academic terms. However, latent thematic analysis did highlight some awareness of these broader issues on the part of some of the participants, and these will be reflected in the discussion that follows.

A number of discourses can be seen to have significance for the spiritual well-being of dependent older people by supporting or undermining reciprocity and interdependency as appropriate life projects for dependent older people. I have chosen to briefly revisit discourses relating to health, disability and citizenship, and then to explore both how discourse relating to governmentality may also have implications for the impeding or promoting of reciprocity, and how the findings reflect this.

Bio-Medical Discourses

The power of ideology and discourse to disseminate assumptions about old age, so that they become widespread and uncritically accepted as 'common-sense', is evident in the work of Gramsci (Nowell-Smith 1998), and is also expressed well by Lawler (2008) as follows:

> The argument being made here is that certain things become 'true', not because of any intrinsic property of the statements themselves, but because they are produced from within authoritative, powerful positions, and the accord with *other* 'truth statements'. They are part of a system of knowledges. As a result they seem to be 'inevitable, unquestionable, necessary', as Ian Hacking puts it (1995:4) (p. 58).

This can be seen to throw some light on the question I found myself asking of the research findings: 'Why would the participants, when asked how they had come to be significantly dependent on others, almost universally refer to themselves as ill'? It had been the case that, when asked about why they had been admitted to a care facility, or assessed as being eligible for care support in the community, all but one of the participants had framed their responses in terms of an incapacitating accident or chronic illness. Of all 14 elders, Muthu had been the only one not to cite a health issue directly, referring instead to 'disengagement from family'. I had prefaced each interview with an exploration of why that individual had come to need help with their daily lives, and each participant had immediately framed their response in terms of illness before I had felt any need to prompt them with the suggestion that this might have been one reason among others. Given the constant drip-feed of ageist messages to which the UK-based participants are likely to have been subjected (through, for example, media images portraying them as decrepit or incompetent (Bowd 2003)—I had expected that at least some of them might have answered that question by referring to themselves as old, even if in a self-deprecatory manner, but none referred to their age, or inferred that they were too old, rather than too sick, to manage alone.

This may be explained in part by the enduring influence of bio-medical discourse in eldercare (Powell 2006), which Leder (cited in Nettleton 2006) suggests can mask other dimensions of experience, such as the spiritual dimension which the findings suggest can often be neglected in the planning or provision of eldercare: 'He argues that biomedicine has focused on the 'body-as-machine', concentrating only on the physical aspects of the body, to the neglect of the mind and the person' (p. 114). I neither have the space, nor feel it necessary for this project, to explore the history of bio-medical discourse, and how old age and illness have come to be so closely associated. However, Pickard's account of the professionalising of geriatric medicine (Pickard 2010) is particularly worthy of mention because it provides food for thought about how differing dominant discourses may have been operating in the two contexts from which my samples were drawn, and supports my argument that the factors which impede or promote reciprocity are dynamic:

> The discourses employed by geriatricians as they attempted to professionalise contributed to the broader problematisation of old age in several ways. As we have noted, the organisational discourses that directly addressed the problem of 'bed blocking' furthered the dualistic approach to old age as 'good' or 'bad'. At the same time, in emphasising the ambiguity surrounding normality and pathology in old age, the conflation of old age generically with ill health was progressed, and the need of all older people for medical intervention or supervision thus implied (pp. 1081–1082).

While this may help to account for the conflation of old age and illness in the public consciousness of UK citizens (and why the UK elders were describing themselves as ill rather than old), old age and illness do not appear to have been associated to the same degree in India, where geriatrics as a medical specialism is still in its relative infancy (Ingle and Nath 2008).

I have therefore considered whether the two concepts might be linked in a similar relationship in the Indian context, and considered that, if geriatrics as a specialism continues to gather momentum in India, the conflation of old age and ill-health may become increasingly apparent there too—perhaps eventually challenging the currently dominant religious (particularly Hindu) discourses that:

- downplay the importance of the physical body in relation to the soul;
- espouse reciprocity; and
- locate it within an ethos of self-responsibility.

Powell (2006) makes reference to a 'professional gaze' (operating akin to the 'medical gaze' described by Foucault 1977) which can be seen to contribute to the perpetuation of a social structure in the UK in which older people are marginalised by locating the power to accommodate, or neglect, the aspirations of older people in the hands of professional gatekeepers of scarce resources (Lymbery 2010). For almost two decades now, unless they have had the private means to bypass state provision, older people's lifestyle choices have been mediated by the judgements of care managers who, in their role as government officers, can be seen to have played a role in maintaining the negative association between dependent old age and reciprocity. While individual workers may espouse empowerment and a respect for unique strengths as part of their value base, the necessity for them to focus on dependent older people's deficits in order to secure funding for care needs, has run contrary to the project of promoting the strengths-based approaches which would value and promote reciprocity (Saleeby 2008; Lamb et al. 2009).

Such insights are helpful in accounting for the impeding of reciprocity as far as UK elders are concerned but in India, where the care of older people does not seem to be mediated by professionals, then other discourses—such as those relating to kinship obligation and moral duty (Datta 2008; Bagga 2008)—may better account for why the Indian elders represented themselves as sick, and remained in residential care facilities even when there was no pressing medical need for them to continue living there.

Regardless of the extent to which bio-medical discourses can be said to account for the conflation of old age and sickness, I would suggest that this association nevertheless has significance in relation to expectations of reciprocity because:

- being sick is taken, in common-sense thinking, to sanction a suspension of obligations to others, (Varul 2010, with reference to the work of Talcott Parsons);
- conceptualising and explaining one's dependency in terms of being sick, rather than old, may be a strategy on the part of dependent older people to disassociate themselves from the negative stereotypes commonly associated with old age; and
- associating oneself with a sick role is perhaps less threatening to a person's sense of identity than accepting the label 'old', because the latter may imply a *permanent* loss of a valued role, whereas the label 'sick' holds some promise of a resumption of usefulness after a period of incapacity. In that sense, then, the findings appear to reflect the premise of biographical disruption (Bury 1982) as possibly accounting for why so many of the participants described their dependency with reference to ill-health, and why some, including Kannamma, Maureen and Eddie, perceived themselves as having the potential to pick up where they had left off in terms of reciprocity, should their health improve.

Testament to the continuing dominance of bio-medical discourses in eldercare is the extent to which the participants in both settings acquiesced to the focus of care being almost exclusively on physical matters, to the neglect of spiritual and ontological needs. As already indicated, even where participants recognised that their need to have a sense of purpose and value was being neglected, they appeared to accept it nevertheless. To have done otherwise would have been to risk being pathologised, as may have been happening in Eddie's case when his comments about the administration of medicine had been ridiculed, and in Ken's when his attempts to teach his carers about household management were ignored.

Disability Discourses

In the Chennai context, disability appeared to be largely understood with reference to a medical model. Evident from informal feedback at a newly emerging disability focus group in Chennai, and in the narratives of the study participants, was that it still tended to be pathologised there as individual impairment, and reliance on family or charitable support appeared to be the expected norm. In the UK, disability studies reflect a challenge to this, in the form of an understanding of disability as a social construction, and of disabled people as in need of empowerment, rather than care (Morris 2004; Barnes and Sheldon 2010). However, where disability is associated with old age, discrimination on the grounds of both age and disability appear to be mutually reinforcing (Zarb 1993), and older disabled people less visible than their younger counterparts in disability activism (Jonson and Larsson 2009). A key part of ageist ideology is the assumption that old age itself can be seen as a form of disability, and where disability co-exists with old age, the disability can be seen to be 'invisibilised'—subsumed under the definition of aged because dominant ageist discourses promote the assumption that to be old is to be disabled anyway—by definition:

> A consequence of equating old age with disability is that it can serve to strengthen the link in people's minds between old age, frailty and dependency. And a further consequence of that process is the assumption that ill-health or disability in old age is caused by the ageing process and therefore has certain inevitability about it which justifies giving it little attention or priority (Thompson 2005, p. 72).

So, while there is a challenge in the UK to discourses which pathologise disability, informed by social models of disability (Oliver and Sapey 2006), *older* people who acquire a disability continue to be relatively marginalised within those more empowering discourses and their disability underpoliticised as a consequence (Barnes and Sheldon 2010). As such, they can be said to be doubly pathologised, in that their marginalisation is legitimised by ageist discourses, but also by disability discourses because of the equating of old age with disability.

The argument that disability is socially constructed through disabling attitudes and environments (Barnes and Mercer 2010) chimes with the earlier discussion of how the participants in this study were, for the most part, isolated from opportunities for intersubjective feedback that they were 'good' people, by a number of attitudinal, financial and ideological barriers. And while discourses tend to operate unnoticed (Gilbert and Powell 2010), some of the participants had shown an awareness of barriers that were being placed in the way of their aims to operate as valued people in their communities. This was evident in:

- Eddie showing an awareness that policy directives associated with risk management were constraining him from operating in what he considered to be useful and appropriate ways;
- Ken being aware that it was not just the attitudes of care staff that were constraining him, but also environmental barriers, such as the restrictive layout of the social club that was integral to his aspiration to be of more use to others; and
- Muthu understanding that his wish to do voluntary work could not be facilitated because no-one who needed the help he could offer ever came to the residential home. He made no mention of the possibility that he could have been enabled to access those opportunities by being helped to *leave* the care home.

In spite of being aware of these disabling processes they, and the majority of their fellow participants, acquiesced to being denied the connectedness which would have held promise for promoting opportunities for reciprocity. In the face of mutually reinforcing ageist and disablist discourses which contributed towards establishing passivity and dependency as norms, this was not a surprising finding.

Governmentality Discourse

Governmentality is a contested concept, but I use it here with reference to how government can be seen to be a complex set of processes operating in different forms, and at different times, to maintain social stability. In the Foucauldian sense it can be seen to refer to a form of self-policing, whereby people contribute to the maintenance of social order by 'governing' themselves (Conway and Crawshaw 2009; Pickard 2010). That is, they can be said to internalise 'rules' about what is expected of them, and live their lives influenced by what dominant discourses promote as the 'right and proper' way to think and behave—in this case, to be 'done to' in old age, rather than 'to engage with' and 'to contribute to'. Given the high prevalence in the findings of accounts of acquiescence to a lack of opportunities for reciprocity, this explanation may have some currency.

It seems that, both in the UK and India, policy making and welfare delivery relating to eldercare is not enacted through government departments alone, but increasingly in alliance with other interests, including:

- the business sector, such as private care homes and domiciliary care agencies;
- non-governmental organisations, especially in India;
- lobbying groups such as user involvement initiatives, and charities such as HelpAge India and Age UK; and
- research bodies such as the Joseph Rowntree Foundation.

Each of these parties can be seen to have different interests and, in terms of the implications for the impeding or promoting of reciprocity, may not all have a respect for spiritual well-being as a guiding principle, or even an interest at all. In the midst of conflicting discourses about whether or not they can consider themselves to be valued citizens, it is not surprising that older people dependent on formal care may experience difficulties maintaining a positive self-image, as was reflected in the frustrations and disappointments expressed in several of the narratives—where a desire to reciprocate was strong, but 'permission' to do so was neither perceived to be in their own hands, nor sought from others.

Several of the Indian elders commented, or implied, that old age continues to be revered, such that older people are still considered there to be 'useful' citizens, regardless of age—and that they were trying to live up to that ideal, even when ill (Subbulakshmi and Muthu, for example). It is possible that this will change in the face of demographic changes which are resulting in an increase in the numbers of frail older people requiring help and a reduction in the availability of, and commitment to, family support (Bagga 2008)—to the effect that older people may become increasingly 'problematised' (Fennell et al. 1988; Pickard 2010). In light of this, if older people come to be increasingly conceptualised and problematised as a burden, rather than revered as an asset, then positive attributes such as Muthu's perceived 'helping tendency' may come to be overshadowed by blanket assumptions that old age and a 'helping tendency' are incompatible.

It seems that the UK elders were also receiving mixed messages about the extent to which they are seen as competent. For example, the personalisation agenda already referred to seems, on the face of it, to have potential for helping people to maintain a positive identity—to 'be me, and stay me'—by devolving care budgets to individuals, so that they can personalise their care provision. By doing so, power can also devolved so that, in having the freedom to address their weaknesses they have the potential to also maximise their strengths. Being offered this option is to receive the message that to be old is not necessarily to be incapable—itself a recognition that the potential for 'usefulness' exists. However, it seems that this potential has yet to be realised in eldercare delivery (Manthorpe and Stevens 2010).

Citizenship Discourses

Much of what has already been discussed in terms of the power of discourse relates to the extent to which older people are considered to *be* citizens of the states in which they reside—that, is to retain the citizenship status they had held prior to becoming dependent on formal care in old age. Discourse that focuses on citizenship would appear to have positive implications for the promoting of reciprocity in the lives and care of elders because citizenship incorporates not only rights, but also responsibilities (Lewis 2004; Barry 2005)—such that it could be argued that reciprocity should be not only valued in, but *expected of,* citizens. A citizenship perspective, then, has the potential to frame old age as, broadly speaking, a time of interdependency, rather than dependency, and therefore to challenge ageist and bio-medical discourses that associate it with decline and burden.

However, as intimated earlier, the potential for citizenship discourses to promote reciprocity in eldercare can be seen to be undermined by challenges to older people's claim to citizenship (Midwinter 1990). User involvement initiatives have played a part in highlighting that old age does not preclude citizenship, and in promoting and facilitating older people's rights as citizens to have their perspectives heard in decision-making arenas (Thornton 1995; Cornes et al. 2008; Iliffe et al. 2010). As such, the potential for reciprocity in eldercare to be promoted would seem to be enhanced by such initiatives, where they exist. But while several of the participants in both samples had declared a willingness to engage in such arenas, any opportunity that was offered was perceived by the participants (and I found myself concurring) as tokenistic, and, as such, a potential impediment in itself.

Conclusion

My aim in this chapter has been to highlight how the findings of this study have been able to address, in part, the relative lack of debate in reciprocity and eldercare studies that relate to the significance of reciprocity for spiritual well-being, and particularly for the spiritual well-being of older people dependent on formal care. The findings have highlighted that meaning making at the level of personal spirituality may be compromised where:

(a) older people are not considered to have a future dimension; and
(b) opportunities for social affirmation are neglected.

My analysis has been innovative in using a phenomenologically grounded conceptual framework which incorporates social space and social time as enlightening concepts, to inform: (a) how dependent older people make sense of the presence, or absence, of reciprocity in their lives, (b) an understanding of how they may be forced into a degraded, disempowering form of reciprocity and (c) an appreciation of the potential of a future orientation to provide a platform on which to build an argument that a recent trend in the theorising of old age which conceptualises it as a positive time of life be extended to the 'fourth age' too. It is my understanding that relating individual meaning making about reciprocity in eldercare to both the environments and relationships in which it flourishes (or otherwise) and the sense in which this meaning making takes place in changing contexts, is innovative in itself. So, too has been my use of PCS analysis to provide a platform for extending our understanding of reciprocity and its significance for the spiritual well-being of older people dependent on formal care. I now turn to a discussion of how my work also has significance for the theorising of old age itself.

References

Antonucci, T., & Jackson, J. (1989). 'Successful ageing and lifecourse reciprocity', in Warnes (1989).
Arber, S., Davidson, K., & Ginn, J. (2003a). 'Changing approaches to gender and later life', in Arber et al. (2003).
Arber, S., Davidson, K., & Ginn, J. (Eds.), (2003b). *Gender and ageing: Changing roles and relationships*. Maidenhead: Open University Press.
Attig, T. (2011). *How we grieve: Relearning the world* (revised edn). New York: Oxford University Press.
Baars, J. (2007). Chronological time and chronological age: Problem of temporal diversity, in Baars and Visser (2007).
Baars, J., & Visser, H. (Eds.), (2007). *Aging and time*. New York: Baywood.
Bagga, A. (2008). 'Gender issues in care giving', in Chatterjee et al. (2008).
Barnes, C., & Mercer, G. (2010). *Exploring disability* (2nd ed.). Cambridge: Polity Press.
Barnes, C., & Sheldon, A. (2010). Disability politics and poverty in a majority world context. *Disability and Society, 25*(7), 771–782.
Barry, B. (2005). *Why social justice matters*. Cambridge: Polity Press.
Beck, U., & Beck-Gernsheim, E. (2001). *Individualization*. London: Sage.
Beel-Bates, C. A., Ingersoll-Dayton, B., & Nelson, E. (2007). Deference as a form of reciprocity among residents in assisted living. *Research on Aging, 29*, 626–643.
Beresford, P., & Croft, S. (2001). Service users' knowledge and the social construction of social work. *Journal of Social Work, 1*(3), 295–316.
Biggs, S. (1999). *The mature imagination: Dynamics of identity in midlife and beyond*. Buckingham: Open University Press.
Boerner, K., & Reinhardt, J. P. (2003). Giving while in need: Support provided by disabled older adults. *Journal of Gerontology, 58B*(5), S297–S304.
Bond, J., & Cabrero, G. R. (2007). 'Health and dependency in later life', in Bond et al. (2007).
Bond, J., Peace, S., Dittman-Kohli, F., & Westerhof, G. (Eds.). (2007). *Ageing in society: European perspectives on gerontology* (3rd ed.). London: Sage.
Boneham, M. (2002). 'Researching ageing in different cultures', in Jamieson and Victor (2002).
Bornat, J., & Bytheway, B. (2010). Perceptions and presentations of living with risk in everyday life. *British Journal of Social Work, 40*, 1118–1134.
Bourdieu, P. (1984). *Distinction: Social critique of the judgement of taste*. Cambridge: Harvard University Press.
Bourdieu, P. (1986). 'The forms of capital'. In Richardson (Ed.), *Handbook of Theory and Research for the Sociology of Education*. Greenwood: Westport.
Bowd, A. D. (2003). Stereotypes of elderly persons in narrative jokes. *Research on Ageing, 25*(3), 22–35.
Bowers, B. J., Fibich, B., & Jacobson, N. (2001). Care-as-service, care-as-relating, care-as-comfort: Understanding nursing home residents' definitions of quality. *The Gerontologist, 41*(4), 539–545.
Breheny, M., & Stevens, C. (2009). "I sort of pay back in my own little way": Managing independence and social connectedness through reciprocity. *Ageing and Society, 29*, 1295–1313.
Buber, M. (1958). *I and Thou* (2nd ed.). London: Continuum.
Bury, M. (1982). Chronic illness as biographical disruption. *Sociology of Health & Illness, 4*, 165–182.
Butler, R. N. (1987). *'Ageism', in the Encyclopedia of aging 22–23*. New York: Springer.
Bytheway, B. (2011). *Unmasking age: The significance of age for social research*. Bristol: The Policy Press.
Bytheway, B., Bacigalupo, V., Bornat, J., Johnson, J., & Spurr, S. (Eds.), (2002). *Understanding care, Welfare and Community*. London: Routledge.

Cann, P., & Dean, M. (Eds.), (2009). *Unequal ageing: The untold story of exclusion in old age.* Bristol: The Policy Press.
Castiglione, D., van Deth, J. W., & Wolleb, G. (Eds.), (2008). *The handbook of social capital.* Oxford: Oxford University Press.
Chakraborti, R. D. (2004). *The greying of India: Population ageing in the context of Asia.* New Delhi, India: Sage.
Chatterjee, D. P. (2008). Oriental disadvantage versus occidental exuberance. *International Sociology, 23*(1), 5–33.
Chatterjee, D. P., Patnaik, P., & Chariar, V. M. (Eds.), (2008). *Discourses on aging and dying.* London: Sage.
Chokkanathan, S. (2009). 'Resources, stressors and psychological distress among older adults in Chennai. India', *Social Science and Medicine, 68*, 243–250.
Conway, S., & Crawshaw, P. (2009). "Healthy senior citizenship" in voluntary and community organisations: A study in governmentality. *Health Sociology Review, 18*, 387–398.
Cook, G. A. (2008). Older people actively reconstruct their life in a care home. *International Journal of Older People Nursing, 3*(4), 270–273.
Coote, A. (2009). 'The uneven dividend: Health and well-being in later life', in Cann and Dean (2009).
Cornes, M., Peardon, J., & Manthorpe, J. (2008). Wise owls and professors: The role of older researchers in the review of the National Services Framework for older people. *Health Expectation, 11*(4), 409–417.
Coyte, M. E., Gilbert, P., & Nicholls, V. (Eds.), (2007). *Spirituality, values and mental health: Jewels for the journey.* London: Jessica Kingsley.
Crossley, N. (1996). *Intersubjectivity: The fabric of social becoming.* London: Sage.
Daniel, S. (1999). The healthy patient: Empowering women in their encounters with the healthcare system. *The American Journal of Clinical Hypnosis', 42*(2), 108–114.
Datta, A. (2008). 'Socio-ethical issues in the existing paradigm of care for the older persons: Emerging challenges and possible responses', in Chatterjee et al. (2008).
de Beauvoir, S. (1977). *Old age.* Harmondsworth: Penguin.
Doka, K. J. (Ed.), (2001). *Disenfranchised grief* (3rd ed.). New York, NY: Lexington.
Farndon, J. (2007). *India booms: The breathtaking development and influence of Modern India.* London: Virgin Books.
Fennell, G., Phillipson, C., & Evers, H. (1988). *The sociology of old age.* Milton Keynes: Open University Press.
Fennema, M., & Tillie, J. (2008). 'Social capital in multicultural societies', in Castiglione et al. (2008).
Fernando, S. (2007). 'Spirituality and mental health across cultures', in Coyte, Gilbert and Nicholls (2007).
Foucault, M. (1977). *Discipline and punish: The birth of the prison.* London: Allen Lane.
Fromm, E. (1955). *The sane society.* New York: Rinehart.
Gadamer, H.-G. (2004). *Truth and method* (3rd ed.). London: Continuum.
Gandhi, K., & Bowers, H. (2008). *Duty and obligation: The invisible glue in services and support.* York: The Joseph Rowntree Foundation.
Gardner, P. J. (2010). 'The public life of older people neighbourhoods and networks' *Dissertation Abstracts International*, Section B, Vol 17 (4-B) p. 2276.
Garey, A. I., Hansen, K. V., Hertz, R., & Macdonald, C. (2002). Care and kinship: An introduction. *Journal of Family Issues, 23*, 703–715.
Garrett, P. M. (2010). Making social work more Habermasian? A rejoinder in the debate on Habermas. *The British Journal of Social Work, 40*(6), 1754–1758.
George, K. L. (2008). 'Paternalistic decisions for the comate and dying aged: A neo-vedantic perspective, in Chatterjee et al. (2008).
Gilbert, T., & Powell, J. L. (2010). Power and social work in the United Kingdom; A Foucauldian excursion. *Journal of Social Work, 10*(1), 3–22.
Gilleard, C., & Higgs, P. (2005). *Contexts of ageing: Class cohort and community.* Cambridge: Polity Press.

References

Gilleard, C., & Higgs, P. (2010). Aging without agency: theorising the fourth age. *Aging and Mental Health, 14*(2), 121–128.

Gilleard, C., & Higgs, P. (2011). Ageing abjection and embodiment in the 4th age. *Journal of Ageing Studies, 25*(2), 135–142.

Goldsmith, M. (2004). 'The stars only shine in the night: The challenge of creative ageing', in Jewell (2004).

Gouldner, A. W. (1960). The norm of reciprocity: a preliminary statement. *American Sociological Review, 25*(2), 161–178.

Gray, A. (2009). The social capital of older people. *Ageing and Society, 29*, 5–31.

Grenier, A., & Hanley, J. (2007). Older women and "frailty": Aged, gendered and embodied resistance. *Current Sociology, 55*, 211–228.

Hague, G., & Mullender, A. (2006). Who listens? The voices of domestic violence survivors in service provision in the UK. *Violence Against Women, 12*(6), 568–587.

Hatton-Yeo, A. (ed.), (2006). Intergenerational programmes: An introduction and examples of practice, Stoke-on-trent, Beth Johnson Foundation

Heidegger, M. (1962 trans.). *Being and time: A translation of Sein und Zeit*, New York: University of New York Press.

Hickey, G., & Kipping, C. (1998). Exploring the concept of user involvement in mental health through a participation continuum. *Journal of Clinical Nursing, 7*, 83–88.

Hill, R. D. (2005). *Positive aging: A guide for mental health professionals and consumers*. London: W.W. Norton.

Hiller, H. H., & Franz, T. M. (2004). New ties, old ties and lost ties: the use of internet in diaspora. *New Media Society, 6*, 731–752.

Hodge, D. R. (2004). Working with Hindu clients in a spiritually sensitive manner. *Social Work, 49*(1), 27–38.

Hudson, J. (2002). 'Community care in the information age', in Bytheway et al. (2002).

Hugman, R. (1991). *Power in the caring professions*. Basingstoke: Palgrave Macmillan.

Iliffe, S., Kharicha, K., Kharicha, D., Swift, C., Goodman, C., & Manthorpe, J. (2010). 'User involvement in the development of a health promotion technology for older people: Findings from the Swish Project', *Social Care in the Community, 18*(2), pp. 147–159.

Ingle, G. K., & Nath, A. (2008). Geriatric health in India: Concerns and solutions. *Indian Journal of Community Medicine, 33*(4), 214–218.

Jacobs, S. (2010). *Hinduism today*. London: Continuum.

Jamieson, A., & Victor, C. R. (Eds.), (2002). *Researching ageing and later life*. Buckingham: Open University Press.

Jewell, A. (2004). 'Nourishing the inner being: A spiritual model', in Jewell (2004).

Jewell, A. (Ed.), (2004b). *Ageing, spirituality and well-being*. London: Jessica Kingsley Publishers.

Jonson, H., & Larsson, A. T. (2009). The exclusion of older people in disability activism and policies: A case of inadvertent ageism? *Journal of Aging Studies, 23*(1), 69–77.

Jordan, B. (2008). *Welfare and well-being: Social value in public policy*. Bristol: The Policy Press.

Knight, T., & Mellor, D. (2007). 'Social inclusion of older adults in care: Is it just a question of providing activities?'*International Journal of Qualitative Studies in Health and Well-being, 2*(2), pp. 74–85.

Komter, A. (2007). Gifts and social relations: The mechanisms of reciprocity. *International Sociology, 22*, 93–106.

Lamb, F. F., Brady, E. M., & Lohman, C. (2009). Lifelong resiliency learning: A strengths-based synergy for gerontological social work. *Journal of Gerontological Social Work, 52*, 713–728.

Lawler, S. (2008). *Identity: Sociological perspectives*. Cambridge: Polity Press.

Lewis, G. (Ed.), (2004). *Citizenship: personal lives and social policy*. Bristol: The Policy Press.

Lin, N. (2008). 'A network theory of social capital', in Castiglione et al. (2008).

Lister, R. (2003). *Citizenship: Feminist perspectives* (2nd ed.). Palgrave Macmillan: Basingstoke.

Lloyd, M. (2008). From service user to VIP: What's in a name? *Journal of Mental Health Training, Education and Practice, 3*(3), 53–54.

Lymbery, M. (2010). A new vision for adult social care? Continuities and change in the care of older people. *Critical Social Policy, 30*(5), 5–26.
Mackinlay, E. (2001). *The spiritual dimension of ageing*. London: Jessica Kingsley Publishers.
Manthorpe, J. (2007). Managing risk in social care in the United Kingdom. *Health, Risk and Society, 9*(3), 237–239.
Manthorpe, J., & Stevens, M. (2010). Increasing care options in the countryside: Developing an understanding of the potential impact of personalization for social work with rural older people. *British Journal of Social Work, 40*(5), 1452–1469.
McMunn, A., Nazroo, J., Wahrendorf, M., Breeze, E., & Zaninotto, P. (2009). Participation in socially-productive activities, reciprocity and well-being in later life: baseline results in England. *Ageing and Society, 29*, 765–782.
Mehta, K. (1997). Cultural scripts and the social integration of older people. *Ageing and Society, 17*(3), 253–275.
Midwinter, E. (1990). An ageing world: The equivocal response. *Ageing and Society, 10*(2), 221–228.
Moriarty, J., & Butt, J. (2004). 'Social support and ethnicity', in Walker and Hennessy (2004).
Morris, J. (2004). Independent living and community care: A disempowering framework. *Disability and Society, 19*(5), 427–442.
Moss, B. (2005). *Religion and spirituality*. Lyme Regis: Russell House Publishing.
Moss, B. (2007). *Values*. Lyme Regis: Russell House Publishing.
Moss, B. (2008). Perspectives. *Well-being e-zine, 4*(2), 2.
Neimeyer, R. A., Harris, D. L., Winokuer, H. R., & Thornton, G. F. (Eds.), (2011). *Grief and bereavement: Bridging research and practice*. London: Routledge.
Neimeyer, R. A., & Sands, D. C. (2011). 'Meaning reconstruction in bereavement: From principles to practice', in Neimeyer et al. (2011).
Nelson, T. D. (Ed.). (2002). *Ageism: Stereotyping and prejudice against older persons*. London: Massachusetts Institute of Technology.
Nettleton, S. (2006). *The sociology of health and illness* (2nd ed.). Cambridge: Polity Press.
Neuberger, J. (2009). 'What does it mean to be old?' in Cann and Dean (2009).
Nowell-Smith, G. (Ed.), (1998). *Antonio Gramsci: Selections from the Prison Notebooks*. London: Lawrence and Wisehart.
Oliver, M., & Sapey, B. (2006). *Social work with disabled people* (3rd ed.) Basingstoke: Palgrave Macmillan.
Park, A., Phillips, M., Clery, E., & Curtice, J. (2010). *British social attitudes survey 2010–2011: Exploring Labour's legacy, the 27th report*. London: Sage.
Peace, S., Holland, C., & Kellaher, L. (2006). *Environment and identity in later life*. Maidenhead: Open University Press.
Pearlin, L. I., & Schooler, C. (1978). The structure of coping. *Journal of Health and Social Behaviour, 22*, 337–356.
Pickard, S. (2010). The 'Good Carer': Moral practices in late modernity. *Sociology, 44*, 471–486.
Powell, J. L. (2006). *Social theory and ageing*. Oxford: Rowman and Littlefield.
Powell, J. L., & Biggs, S. (2000). Managing old age: The disciplinary web of power, surveillance and normalisation. *Journal of Aging and Identity, 5*(1), 3–13.
Putnam, R. D. (2000). *Bowling alone: The collapse and revival of american community*. New York: Simon and Schuster.
Renzenbrink, I. (2004). Home is where the heart is. *Illness, Crisis and Loss, 12*(1), 63–74.
Renzenbrink, I. (2010). *Fluttering on fences: Stories of loss and change*. Canada, Lakeside Education and Training: Saskatoon.
Richardson, J. G. (Ed.), (1986). *Handbook of Theory and Research for the Sociology of Education*. Greenwood: Westport.
Rowbotham, S. (1973). *Woman's consciousness*. Harmondsworth, Penguin: Man's World.
Rubenstein, D. (2001). *Culture, structure and agency; towards a truly multidimensional society*. London: Sage.

References

Saleeby, D. (2008). *The strengths perspective in social work practice* (5th ed.). London: Rugby, Pearson Education.
Sartre, J.-P. (1963). *Search for a method*. New York: Vintage.
Saunders, C. (2002). 'The philosophy of hospice', in Thompson (2002).
Swain, J., Finklelstein, V., French, S., & Oliver, M. (Eds.), (1993). *Disabling barriers: Enabling environments* (2nd ed.). London: Sage.
Tanner, D. (2010). *Managing the ageing experience: Learning from older people*. Bristol: The Policy Press.
Thompson, N. (1998). The ontology of ageing. *The British Journal of Social Work, 28*(5), 695–707.
Thompson, N. (Ed.), (2002). *Loss and grief*. Basingstoke: Palgrave Macmillan.
Thompson, S. (2005). *Age discrimination*. Lyme Regis: Russell House Publishing.
Thompson, N. (2007). *Power and empowerment*. Lyme Regis: Russell House Publishing.
Thompson, S. (2007). Spirituality and old age. *Illness, Crisis and Loss, 15*(2), 169–181.
Thompson, N. (2011a). *Promoting equality: Working with diversity and difference* (3rd ed.). Basingstoke: Palgrave Macmillan.
Thompson, N. (2011b). *Crisis intervention*. Lyme Regis: Russell House Publishing.
Thompson, N., & Thompson, S. (2001). Empowering older people: Beyond the care model. *Journal of Social Work, 1*(1), 61–76.
Thompson, S., & Thompson, N. (2008). *The critically reflective practitioner*. Basingstoke: Palgrave Macmillan.
Thornton, P. (1995). *Having a say in change*. Joseph Rowntree Foundation: York.
Timonen, V. (2008). *Ageing societies: A comparative introduction*. Maidenhead: Open University Press.
Twigg, J. (2004). The body, gender and age: Feminist insights in social gerontology. *Journal of Ageing Studies, 18*, 59–74.
Uehara, E. S. (1995). Reciprocity reconsidered: Gouldner's "moral norm of reciprocity" and social support. *Journal of Social and Personal Relationships, 2*, 483–502.
Ungerson, C. (2000). Cash in care, in Harrington Meyer (2000).
Varma, P. K. (2005). *Being Indian: The truth about why the twenty-first century will be India's*. London: Penguin.
Varul, M. Z. (2010). Talcott Parsons, the sick role and chronic illness. *Body and Society, 16*, 72–94.
Victor, C., Scambler, S., & Bond, J. (2009). *The social world of older people: Understanding loneliness and social isolation in later life*. Maidenhead: Open University Press.
Walker, A., & Hagan Hennessy, C. (Eds.), (2004). *Growing older: Quality of life in old age*. Maidenhead: Open University Press.
Warnes, A. M. (Ed.), (1989). *Human ageing and later life: Multidisciplinary perspectives*. London: Edward Arnold.
Webster, J., & Whitlock, M. (2003). Patient or person? *Nursing Older People, 15*(5), 38–39.
Westerhof, G. J., & Tulle, E. (2007). 'Meanings of ageing and old age: Discursive contexts, social attitudes and personal identities', in Bond et al. (2007).
Wilson, G. (2000). *Understanding old age: Critical and global perspectives*. London: Sage.
Wiseman, D. (2011). *A "four nations" perspective on rights, responsibilities, risk and regulation in adult social care*. York: Joseph Rowntree Foundation.
Zarb, G. (1993). 'The dual experience of ageing with a disability', in Swain et al. (1993).

Chapter 7
The Significance of the Findings for the Theorising of Old Age

Introduction

In Chap. 6 I focused on the significance of the findings of this study for highlighting that a phenomenological perspective on reciprocity can illuminate understanding of it as an aspect of spiritual well-being and, in Chap. 8, I discuss how this can inform eldercare practice across a range of disciplines. However, in addition to how this study extends the literature bases relating to reciprocity and eldercare, it can also be seen to have broader significance for the theorising of old age itself. This can be seen to apply in the following senses by:

1. suggesting that a focus on the interaction between agency, culture and structure transcends postmodern and post-structural theories that have focused more specifically on culture to the relative neglect of agency. That is, in theorising old age by drawing on PCS analysis, I propose an approach which offers coherence without underplaying uniqueness and diversity, and accessibility without losing sight of complexity;
2. contributing to the challenge to logocentricity in the theorising of old age by raising the profile of meaning making and subjective interpretation;
3. highlighting epistemological concerns about the underplaying of dependent older people's role in knowledge production; and
4. illustrating problems inherent in deficit models of ageing, especially in terms of the neglect of aspiration for the future.

I turn now to the first of these.

Coherence

As I have already alluded to in Chap. 3, in drawing on PCS analysis as a framework for understanding the significance of reciprocity I am addressing a concern that, in the quest to deconstruct ageing, a sense of theoretical coherence may have become lost in the process. I would reiterate here that, in seeking to challenge the tendency for postmodernism to reject a focus on coherence, I am not advocating an overarching, 'grand' theorising of ageing that reifies it as a disembodied and undifferentiated concept which loses sight of the diversity and humanity within it. Rather, in proposing that reciprocity in old age can be better understood with reference to a theoretical framework that integrates, rather than polarises, structure and agency, I critique those theoretical approaches that are grounded in the assumption that one is more significant than the other for understanding human experience.

As I have argued, purely psychological theories, for example, have attracted criticism from a sociological point of view for their tendency to neglect social context (Stones 2005; Thompson 2010). At the other extreme, some sociological theories, where they focus exclusively on structure, have attracted criticism for neglecting human agency. This emphasis on structure is something that the post-structuralist movement has sought to address through a focus on the significance of discourse and language (Foucault 1977) but, as with postmodernist theorising, it too tends to neglect human agency (Archer 2000).

I have already acknowledged the influence of Giddens' structuration thesis (Giddens 2009) on this study, in that he critiques approaches that focus on *either* structure or agency by proposing that it is the dialectical interaction *between* structure and agency that has better explanatory power than a focus on either one or the other. In searching for a holistic and integrated theoretical framework for making sense of reciprocity in the lives of dependent older people I have recognised the potential for PCS analysis to build on the concept of a dialectic between structure and agency by highlighting two interconnected dialectical process. Below, I briefly revisit the main tenets of PCS analysis before commenting on how the findings have reinforced my premise that it offers a coherence lacking in theoretical approaches which focus on deconstructing ageing.

PCS analysis (Thompson 2011) brings to the fore the interaction between three levels of analysis of the social world: personal, cultural and structural—interaction that he describes as a double dialectic. The first dialectic operates between the personal level (individual experience, thoughts and actions) and the cultural level (that of shared understandings of what can be taken for granted; cultural assumptions—in this case, about what old age is assumed to mean). The dialectic is so named because the interaction is two-way in that:

1. individuals (personally, but on a collective basis) can be said to experience life in old age as influenced by shared understandings of what it means to be old, and powerful discourses that (in the UK, for example), link old age with deficit, ill-health, all-encompassing dependency and so on; but
2. as individuals (collectively) are also able to exert an influence on those shared understandings of old age at the cultural level by either reinforcing them as representative of the norm (and thereby sustaining them), or challenging them as inappropriate. In this sense, their potential power can be described as 'power with'—a term that Thompson (2007a, b) attributes to Rowlands (1998) when he highlights its sociological aspect:

> This is an important concept in terms of empowerment, as it helps to establish that working together collaboratively can be a useful way forward… people in disadvantaged positions can work collectively to pursue their goals—that is, they can find that their concerted, collective power is much greater than isolated efforts to bring about change (strength in unity). It is therefore important to think of empowerment as more than simply an individual or personal matter (p. 16).

However, that reinforcement may only be effective if the challenge is visible which, in this case, can prove difficult, given that dependent older people may be hidden from view through processes of marginalisation within communities (Phillipson 2007; Gilleard and Higgs 2010).

The interaction between the cultural level and the structural level then forms the second dialectic, which also operates on the basis of mutual reinforcement. It is premised that cultural assumptions both reflect and reinforce, structural arrangements. In the case of ageing in the UK, for example, the fact that age has significance as a basis for social division (such that older people are treated as less deserving of resources—income, healthcare, transport and education, for example—and prestige), serves to reinforce the shared assumption at the cultural level that that old age is about inferiority, marginalisation, decline, need and so on. And for as long as those assumptions remain unchallenged, they help to reinforce the legitimacy of age as a basis for social division and an unequal distribution of resources, life chances and prestige.

Did the findings reflect the operating of a double dialectic, which would support my proposition that a sensitising theory such as this could help lay the foundations for a development in the theorising of ageing? I would argue that they did, given that differences between how dependent old age is understood at the C level in the UK and India, and the extent to which age is considered to be a significant basis for social division in both societies, appeared to be reflected in the personal narratives of the two groups of participants.

For example, as is evident from the literature review, those from the UK sample had been experiencing dependency in their old age at a personal level in the context of:

1. age being considered significant as a basis for social division (the S level); and
2. the conceptualising of frail older people as only being in need of looking after being a pervasive and enduring one (at the C level).

In general, their narratives reflected that the shared understandings of old age at the C level had had an impact on their aspirations to remain 'useful', and on their experience of the facilitating of reciprocity in their lives—as was commented in Chap. 6 with regard to the significance of the findings for their spiritual well-being.

I return here to one of those narratives, in order to demonstrate the capacity of PCS analysis to provide theoretical coherence.

It can be seen from Fig. 7.1 that, where Eddie (and dependent older people like him) have the competence, but not the opportunity, to reciprocate, the shared assumption of dependency at the cultural level will continue to be reinforced. And, while it remains unchallenged, it will continue to both reinforce the legitimacy of age as a social division, and negatively affect individual lived experience for those who see themselves as capable of interdependency.

Megan's experience of reciprocity also took place within the same structural and cultural contexts as did Eddie's, but PCS analysis highlights how a change in how old age as conceptualised at the C level can effect change at the P level, and establish a new and mutually reinforcing dialectic between the P and C levels (see Fig. 7.2). The organisation Age UK, in challenging the conceptualisation of old age as burden (through their recognition of the potential for dependent older people themselves to challenge that negative stereotype by example) have very publicly demonstrated that reciprocity and dependency are compatible. For Megan, this challenge to shared assumptions about dependent old age at the C level facilitated her being able to adopt such a role and thereby continue to have what she perceived to be a 'useful' role—akin to the advocacy role that had defined her life since childhood.

Where physically frail but otherwise competent people like herself are 'allowed' to make that usefulness to society visible, the potential for negative stereotypes to be challenged would seem to be strengthened. As a consequence, if dependent old age becomes increasingly understood in terms of strengths rather than deficits (as is happening in terms of the theorising of the 'third age' as positive ageing—Gilleard and Higgs 2005), then the potential exists for the legitimacy of a social structure which discriminates on the grounds of age to be eroded, and thereby less able to exert an influence at the C level. In this sense, drawing on PCS analysis to understand dependent old age not only offers coherence but also, in its inherent dynamism, takes into account the relationship between social space and social time.

The narratives of those experiencing old age in Chennai, a different cultural and structural context, also reflected the operating of a double dialectic with

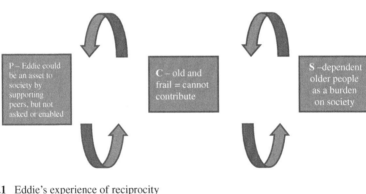

Fig. 7.1 Eddie's experience of reciprocity

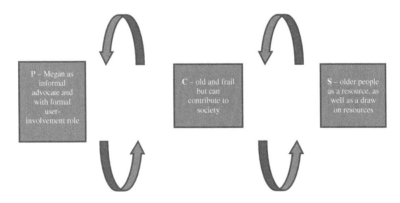

Fig. 7.2 Megan's experience of reciprocity

shared cultural understandings of old age (and therefore, expectations of older people in terms of reciprocity) at the core. And, as can be seen from the literature review, those Indian elders recounting their aspirations and experiences at a personal level were doing so in the context of:

1. age *not* currently being considered significant as a basis for social division (the S level); and
2. the conceptualising of older people as valued and revered remaining relatively pervasive and enduring, though reportedly this is beginning to be eroded in some parts of the subcontinent, especially in urban communities (Datta 2008). Nevertheless, it was one that still appeared to underpin the meaning making of the participants and their support networks at the C level, in Chennai at least.

As with the UK elders, their narratives in general reflected that shared understandings of old age at the C level had had an impact on their aspirations for, and the facilitation of, reciprocity in their lives (Fig. 7.3).

However, given the narratives of the Indian participants, and the reported rise in the number of eldercare institutions (at least in Chennai), changes at the C level may have consequences for the way Indian society is structured in the future, and for the lived experiences of individuals—see Fig. 7.4.

While Lakshmi had been able to continue reciprocating while in residential care (through teaching), this appears to have been an usual situation which most elders in her situation would not have been able to take advantage of. The other participants' experiences appear to have been more typical of life for older people in the residential care facilities that are beginning to be provided as a resource to fill the gap in eldercare caused by the decline in family care support (Mccabe 2006). It can be argued (see Fig. 7.4) that changes at the C level (old age being increasingly associated with formal rather than family support, and characterised by dependency rather than interdependency) have been instrumental in limiting life-enhancing opportunities for reciprocity, and may account, at least in part, for the expressions of spiritual diminishment that emerged from the findings.

These examples demonstrate that PCS analysis constitutes a form of sensitising theory which challenges essentialism by providing a framework which helps make sense of the complexity and fluidity of social life—in Sibeon's terms, one which can: 'equip us with ways of thinking about the world' (Sibeon 1996, p. 4). And, while helping to 'unpack' reciprocity as a particular aspect of the ageing experience, old age is not deconstructed in the process.

Meaningfulness

The linking thread between PCS analysis as significant for the theorising of old age, and the phenomenologically oriented theoretical framework I have proposed, is that of meaning making.

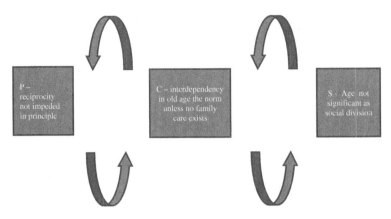

Fig. 7.3 Current Indian elders' experience of reciprocity

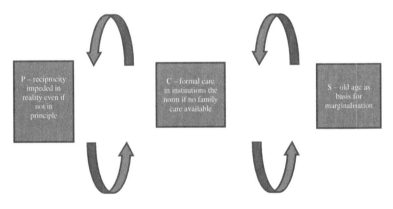

Fig. 7.4 Indian elders' potential experience of reciprocity

Spirituality, an element of my four-part phenomenological analytical framework, can be seen to operate at the P level, while discussion about the significance of discourse for an understanding of the impeding or promoting of reciprocity operates at the C level of shared understandings. And, where I refer, in terms of PCS analysis to the implications for older people of being defined as different and 'less than' at the S level, this can also be seen to have consequences for meaning making—that is, how people are defined and perceived by others, and how those positive or negative perceptions are internalised—such that identity can be said to have social, structural roots (Lawler 2008). It can therefore be argued that, in the sense that the PCS framework incorporates meaning making (and the different levels at which it takes place) it is consistent with the phenomenological framework I have proposed.

The insights that have emerged from the participants' narratives have demonstrated how the theorising of old age neglects a very significant dimension if it does not take account of the meaning making that operates from an 'insider' perspective. Taking on board the proposition that experiences of social phenomena are necessarily subjective (Rorty 2007) and though there were some elements of similarity in the participants' accounts of the impeding or promoting of reciprocity in their lives—they were all unique interpretations because they were made sense of in the context of differing past experiences and future aspirations, even where the structural and cultural contexts were the same. That is, they reflected Heidegger's concept of 'thrownness' (1962) and the interpretation that Thompson (1998) offers of Sartre's regressive-progressive method, in the sense that, apart from Doris, all were able to articulate that their future aspirations, as well as their past experiences, were feeding into their understanding of their old age. That the findings, in a number of cases, reflected a mismatch between assumptions at the C level about how older people do and should live their lives, and how older people *want* to live their lives, indicates that such a theoretical perspective has value if the challenge of theory is to better understand social life and, in this case, old age as part of that life.

What emerged most noticeably from the narratives of those who participated in this study was that their spiritual dimension—what was significant to them in terms of how they saw themselves and who they aspired to be—was often at odds with shared assumptions about old age at the cultural level. This can be seen in Megan's assertion, when reference was made to those activities typically assumed to be what older people like to do, that she did not want to do *'other people's things'*.

This lack of recognition of, and respect for, individual journeys through life was evident in some of the Indian elders' accounts too—where changes in how old age is conceptualised, and eldercare provided, appear to be contributing to a situation where opportunities for reciprocity that match individual aspiration are becoming limited because the significance of connectedness (a pre-requisite for meaningful reciprocity) in their lives is being overlooked in the interests of providing safe, but often isolated, care communities.

The findings also reflected how the participants were trying to make sense of how they were perceived by others—their interpretation of what is made of them by others (Lawler 2008). Harold, for example, referred to the tendency for carers to take over—to treat him as if he were unable to do *anything*—as evident in his comment about the potential for him to be 'smothered'. He had been self-effacing during the interview about what he had been able to give to others throughout his life in general, and had described himself in his old age as 'past his sell-by date', despite being a competent individual with much to offer others. This does not seem surprising if it is considered that, in his state of dependency on formal care, his only source of intersubjective feedback had been limited to his care team who, informed by *their* own perceptions of old age and *their* meaning making in terms of what their role should be, reflected back to him an image informed by weaknesses rather than strengths.

The meaning attached to being able to reciprocate was perhaps more explicit in Ken's narrative, given that he had been able to identify a distinct change in others' perception of him as competent. This had jarred with his own meaning making in the sense that, while his carers saw him as someone to be 'looked after' (and to a schedule and standards imposed by them), he perceived himself differently—as someone compromised by illness and disability, but still keen to fulfil a useful role, if only it could be facilitated.

Muthu's 'nobody asks me' comment when asked about opportunities to reciprocate in his situation also speaks of a mismatch between his meaning making and that associated with old age at the cultural level, and the implications of that for eldercare if dominant discourses of age as decline and deficit render all others silent and invisible.

Where the theorising of old age lacks a meaning making perspective, the implications of being isolated from community life (the reality for many of those dependent on formal care support) may also not be well understood. It has been argued that the vulnerability which informs cultural assumptions about old age is largely socially constructed (Brown 2002; Martin 2007). That is, where a society is structured in such a way that dependent older people are already marginalised economically and ideologically, providing care support that further restricts opportunities for connectedness is likely to compound that marginalisation, thereby making it difficult to conduct their lives in ways other than as vulnerable. And where vulnerability is heightened in the cultural consciousness, what they *could* contribute is overlooked, with the outcome that they tend then to be responded to as if vulnerability were the only aspect of their being—again, the usefulness of the PCS framework is apparent.

I referred in Chap. 3 to the concept of 'acts of resistance' (Grenier and Hanley 2007) whereby the perception of dependent old age as being characterised by an *all-encompassing* dependency can be seen to be challenged through a refusal, in particular care circumstances, to be 'done to'—a refusal to be grateful for care given, for example. While some of the participants in my study articulated how they did (or would have liked to) express their resistance to being denied a useful role, others may not have articulated it as such, but may nevertheless have been doing so in a less obvious way by reframing their acceptance of what they perceived as their lot as an act of reciprocity in itself—a gift to make life easier for others, as expressed in in Maureen's comment about 'going with the rules'. Though this may, on the face of it, have been an act of agency by them, I would argue that it constituted a 'degraded' reciprocity in light of the pressures exerted at the C and S levels for them to conform to expectations that they *should* act in that way.

My point is, that in each narrative, each participant was articulating how they were interpreting their old age, in their particular circumstance, in their particular cultural and structural contexts. Yet in terms of the spiritual dimension of people's lives for which this is significant, it seems that individual meaning making in old age is not well understood (Jewell 2004a, b; Thompson 2007; Pickard 2010). The findings from this study help to address this gap in the theorising of old age in a way that goes beyond the implications for individual well-being by locating it in the meaning making that operates at the C level in terms of what is expected of older people—meaning making which itself emerges from, and reinforces the legitimacy or otherwise of age as a basis for social division at the S level.

One of the most significant aspects of the findings for the theorising of old age, then, is their demonstration that, from a phenomenological perspective, there are multiple understandings of it. From a Heideggerian perspective in particular, and with reference to the concept of 'being-in-the-world' (1962), each participant's unique account of their old age can be said to *constitute* old age, in that their existence in old age cannot be separated from their interpretation of it. To return to the commentary on Heidegger's perspective by Sherrat (2006): 'not only is understanding a crucial feature of our existence—but it *is* our existence' (p. 80). To theorise old age without recognising dependent older people as self-interpreting beings, and each interpretation of their lived experience as a form of theorising, raises questions about epistemological validity, to which I now turn.

Silenced Voices? Reciprocity and the Co-construction of Knowledge

While dependent older people are not entirely absent from research development and from decision-making forums (Cornes et al. 2008; Iliffe et al. 2010) I would agree with Peace (2002) that they remain under-represented in the field of knowledge production. I have made explicit how my work highlights the value of multiple perspectives, despite the power of dominant discourses to suggest that some 'truths' are more valid than others (Foucault 1977; Lawler 2008) and which call into question why one version of how old age should be lived out should shape how it is allowed to be lived out, while other versions are either unheard or given less credence.

My findings, in contradicting taken-for-granted assumptions, have demonstrated the value of recognising multiple perspectives. Take, for example, the assumption in much of eldercare provision, and as highlighted in earlier chapters, that old age is defined by deficit and the need for protection, such that dependent older people are assumed to need only:

1. to have their physical needs met;
2. the provision of 'activity' to help slow down atrophy; and
3. some entertainment to help prevent boredom.

What appears to be lacking in such a perspective is that, though there may be some recognition of the fact that dependent older people remain social beings, organised interaction with others tends to be limited to entertainment purposes (Clow and Aitchinson 2009), such that people are 'brought in' to fulfil that purpose. This was reflected in the findings as an understanding of social events and activity sessions as either being for entertainment (Venkatraman, Vadivu), or supposedly to keep their hands and minds busy (Edna, Kannamma, Subbulakshmi, Maureen, and Doris).

For some of the participants, potting plants and making trinkets may have relieved boredom, as was evident in Doris's account, but achieved little more than that.

From the perspectives of Muthu, Doris, Kannamma, Subbulakshmi, Eddie. Ken and Harold, the aspiration to engage with other people held more significance—it being their opportunity to connect *meaningfully* with others, in ways that had previously made them feel that they had something to contribute to the world. This begs the question—if that was *their* 'truth' about their old age—*their* understanding of what it is—then why is it that *their* interpretation of old age as being a time when dependency and reciprocity can co-exist is not a dominant one that informs shared understandings of old age at the cultural level?

This may be due in part to epistemological questions about what constitutes theory in general (Thompson 2010) and the theorising of old age in particular (Gubrium and Wallace 1990). If it is understood in a broad sense as the production of knowledge that helps make sense of things, then the conceptualising of older people as co-constructors of knowledge has validity in terms of the role they can play in the theorising of old age through what Gadamer (2004) refers to as the 'fusion of horizons'—that is, the realisation that differing perspectives exist and that the interplay between perspectives has the potential to develop new insights, and new ways of looking at the world.

The findings indicated that, not only were all of the participants able to offer insights drawn from their lived experience that were illuminating for this particular study, but also that many of them expressed a willingness to engage in similar roles again. Furthermore, the findings indicated that they saw such a role as an act of reciprocity which could countereact their own feelings of despair at having to lead an unfulfilled life. And more than that, they appreciated that their insights could also contribute to the welfare of their peers and future cohorts of older people by stimulating debate in the eldercare field. This had been implicit in their signing of the consent forms which spelled this out (Appendix 3) but Venkatraman, for example, explicitly thanked me for the opportunity to function in that role.

When planning the study I had always intended to propose the interviews to the participants as an opportunity to help develop the knowledge base of eldercare. That this opportunity was readily taken up, appears to reflect that what Thompson (2010) refers to as 'wonder' was still present in their lives, and underpinned their willingness to contribute to the well-being of others through their part in the research endeavour:

> This [wonder] is about being captivated by the marvels of the world in which we live. It is about learning, growth and development ... these are factors that can be subdued or even removed altogether by people's social or personal circumstances (p. 221).

In addition, I would suggest that the findings have constituted a critique of theorising that is wrapped in mystique (Rorty 1999; Thompson 2010), making it inaccessible to everyone but academics. While arguing for the theorising of reciprocity to recognise and address its complexity, I would also support, to an extent, a demystifying of theory relating to old age that would more readily draw practitioners into theoretical debates that might inform, and be informed by, their work. Thompson (2000) refers to 'the practitioner as theorist' (p. 135) but, in co-producing knowledge that contributes to a better understanding of old age, those on the receiving end of care support can arguably also be considered as theorists, or at least co-constructors of theory.

In the course of my research I was privileged to hear 16 accounts of what had made people feel valued earlier in their lives, and whether this was impeded or promoted in their current circumstances. Given that the findings have suggested that participants had not been asked about what was important to them, or whether they had aspirations for the future, it is disheartening to realise that I may have been the only person to have heard that a sense of usefulness continued to be very important to them, and that most of them felt spiritually diminished where it was lacking in their lives. Yet how else might those evaluating whether eldercare provision holistically and adequately meets *all* the needs of those on the receiving end of it, if not through hearing those perspectives? There is already evidence to suggest that older people are beginning to be recognised as partners in the development of theory (Peace 2002; Ray 2007). Though few appear to have recognised the potential for those participants in my study to function as such, by informing theory and practice beyond their immediate environments, they had all demonstrated that they had:

1. something important to say which either reinforced or challenged existing knowledge;
2. the ability to reflect on their experiences; and, for a number of them;
3. aspirations to continue reciprocating in a similar fashion as a way to counteract their frustration at not being considered 'givers' as well as 'takers'.

As such, I would suggest that what was apparent was not a deficit on the part of the older people, but on the part of care systems that did not address attitudinal and physical barriers to enabling them to play a useful role by contributing to knowledge production. And, where older people's role in knowledge production is not facilitated, this can be seen to offer a telling comment on not only the value accorded older people, but on the validity of phenomenological enquiry itself.

Reciprocity and Resilience: Challenging the Stereotype of Deficit

Chief amongst my aims has been to explore reciprocity as a key concept in providing a challenge to theorising that assumes decline and deficit in old age, and the discourses that sustain it. In highlighting that older people have the ability, and the aspiration, to be useful members of society—even where their health has become compromised to the extent that they need to rely on others for formal care support—the findings have indicated that the positive ageing agenda has relevance for those elders in their '4th age' as well as for those in their younger and less dependent old age.

As indicated in Chap. 3, reciprocity in general has long been researched, and reciprocity in eldercare is not an entirely new concept or research focus. Already, reciprocity has been associated with positive well-being in old age (Moriarty and Butt 2004; McMunn et al. 2009), and useful and enlightening insights have emerged from studies that have contributed to challenging deficit models by highlighting dependent older people's strengths (Cook 2008). The notion of 'care-as-relating' relationships (Bowers et al. 2001) to which I have already referred implies 'hidden' strengths, not always recognised by practitioners unless they look beyond their present circumstances: 'aides were acknowledging resident selves other than those related to old age, illness and disability (p. 542). Though no-one had reflected an uncaring attitude *per se* in the narratives of those in my study, 'care-as-relating' did not appear to be reflected either. All but a few of the participants had expressed disappointment at being perceived only as 'people who can't', rather than as 'people who can', and that their strengths, and aspirations to use them for the benefit of others, were subverted by judgements of their competence that were made by those whose opinion carried more weight than their own—nurses, doctors, social workers and so on.

In the following three instances (see Figs. 7.5, 7.6 and 7.7) it can be seen that the judgements informed by deficit/protection models were the ones that prevailed, despite the older person in question not being totally compromised by ill-health or disability, and having the moral and legal right to take risks anyway.

Significant steps have already been made in challenging entrenched assumptions that link old age necessarily with deficit, most pertinently for this study in terms of older people's involvement in education (Luppi 2009; Hafford-Letchfield 2010), volunteering (Morrow-Howell et al. 2005; Larkin et al. 2005) and mastery over one's affairs (Beel-Bates et al. 2007; Chokkanathan 2009). The findings of this study reflected a willingness to be involved in both education and voluntary work which was not generally facilitated where the participants lived in residential settings—either because those who provided the care support were not aware of the residents' aspirations to reciprocate, or because they considered them to be incapable of reciprocating in that way. Even though modern technologies offer a number of avenues for reciprocating even when seriously compromised in terms of health or mobility, none were in evidence. This suggests that the challenge to deficit models of old age has a way to go yet if dependent older people are to attract the same degree of innovative care planning as, for example, younger people with learning difficulties are attracting through initiatives such as the personalisation agenda (Williams et al. 2009; Harper 2010).

I have made reference throughout to the need for connectedness in order to generate the intersubjective feedback that is required for:

1. the maintenance of self-esteem and self-worth; and
2. for the building up of reserves of social capital that help to generate resilience and make visible their strengths, which itself has the potential to challenge models of ageing that focus on weakness only.

The findings have reflected that, apart from Megan, who took the initiative to keep herself involved in community life, any social capital that the remaining participants had accrued could best be described as 'bonding' social capital (Putnam 2000). That is, the relationships they had been able to forge had been restricted to those in circumstances similar to their own, and similarly defined in terms of incapacity and deficit. Given that opportunities to foster relationships with people

Reciprocity and Resilience: Challenging the Stereotype of Deficit

> **Muthu's perspective**
>
> That his need for food, shelter and company are being met but he feels downhearted at not being able to help others to the extent he would wish. He is: a) concerned that other people in need are losing out; and b) worried that not being able to reciprocate may have detrimental consequences in terms of his obligation as a practising Hindu.

This version shapes Muthu's ← lived experience

> **Practitioner perspective**
>
> Because Muthu is uncomplaining, it is presumed that he is happy in the supportive, family-style atmosphere they have created. As all of his needs are being catered for, there is no need for any further action.

Fig. 7.5 Muthu's perspective versus practitioner perspective

> **Ken's perspective**
>
> He accepts his limitations but still wants to be involved in contributing to family and community life in ways he has always done, and which have informed his own opinion of himself as the person who people come to when they need practical advice. Though he realises that he has little time left to live, he wants it to be meaningful life as defined by him.

This version shapes Ken's ← lived experience

> **Practitioner perspective**
>
> As Ken is physically unwell and terminally ill, every effort needs to be made to ensure that, while his physical needs may present as paramount, his social needs must not be neglected. While he is at the daycare facilities provided in response to the latter, he needs to be kept busy and entertained in the interests of his mental health, and so is urged to take part in the activities they have provided.

Fig. 7.6 Ken's perspective versus practitioner perspective

outside of that group (thereby constituting 'bridging' social capital) were not facilitated for either the Indian or UK elders, opportunities to make their strengths visible, and to 'break out of the stereotype' were not in evidence. This may have been because they were not thought to have any strengths, which would support the

> **Eddie's perspective**
>
> His whole life has been characterised by risk taking in a number of different contexts, and he feels patronised by not being allowed to take risks in his current situation. While recognising that his own health is compromised, he still feels that he could play a part in addressing the mental health needs of others and, in doing so, address his own need to feel that he still has a purpose in life.

This version shapes

Eddie's lived ←

experience

> **Practitioner perspective**
>
> As an older person defined as being in need of formal care provision, Eddie needs, above all, to be protected from harm. It is recognised that he has psychological, emotional and social needs, but priority is given to protection from physical harm where risk management strategies articulate with care planning.

Fig. 7.7 Eddie's perspective versus practitioner perspective

thesis that strengths perspectives in the human services are not readily associated with dependent older people (Lamb et al. 2009).

In a similar vein, that dependent older people in the UK appear to be less incorporated than others into the personalisation agenda (which offers a degree of mastery over one's life and care support—Lymbery 2010) may also be premised on the assumption that the concepts of dependent old age and strengths are taken to be incompatible. As the findings have shown, this was not the case and they can, indeed, be seen to be compatible.

For the impeding of reciprocity to attract attention, it has to be identified as a problem to be addressed but, if the dominant assumption at the cultural level is that it is acceptable to neglect the significance of reciprocity to older people dependent on formal care, then it is unlikely to be addressed because of the power of ageist discourses operating at that level to portray that 'take' on eldercare as *the* truth—the *only* way to see the situation. In exploring what dependent older people themselves considered to be their continuing strengths, I consider my research to be building on the foundations that *they* have built, to work towards reframing old age at the C level as a time of both strengths and deficits, as indeed is any other stage of life.

Conclusion

I have argued the case for reciprocity to have greater prominence in social theory relating to ageing because where it is highlighted in debate, it has the potential to challenge entrenched and influential deficit models which undermine the strengths that dependent older people retain, or can develop, despite being compromised in some way by illness or disability. In demonstrating a mismatch between the participants' perceptions of old age as usefulness and growth, and those of their care providers as helplessness and stagnation or decline, the findings have highlighted the significance of meaning making for:

1. supporting the argument that the understanding of old age is contested (Townsend 1981; Westerhof and Tulle 2007); and
2. arguing that the validity of different perspectives need to be taken into account if the theorising of ageing is not to be ego- or ethno-centric.

In demonstrating that the significance of reciprocity in eldercare differs both across cultural contexts and time, drawing on the PCS framework to analyse the findings has highlighted that theorising which does not offer coherence and dynamism cannot adequately capture the experiencing of old age from a sociological perspective. Furthermore, in drawing on phenomenological insights to highlight the competence of older people dependent on formal care to contribute a neglected perspective to the theorising of old age, the potential for the findings of this study to inform eldercare practice is therefore also highlighted. It is to their significance for practice, and phronesis, or practice wisdom, that I now turn.

References

Afshar, H. (Ed.), (1998). *Women and empowerment: Illustrations from the third world*. Basingstoke: Macmillan.
Archer, M. S. (2000). *Being human: The problem of agency*. Cambridge: Cambridge University Press.
Beel-Bates, C. A., Ingersoll-Dayton, B., & Nelson, E. (2007). Deference as a form of reciprocity among residents in assisted living. *Research on Aging, 29*, 626–643.
Bernard, M., & Scharf, T. (Eds.), (2007). *Critical perspectives on aging societies*. Bristol: The Policy Press.
Bond, J., Peace, S., Dittman-Kohli, F., & Westerhof, G. (Eds.), (2007). *Ageing in society: European perspectives on gerontology* (3rd ed.). London: Sage.
Bowers, B. J., Fibich, B., & Jacobson, N. (2001). Care-as-service, care-as-relating, care-as-comfort: Understanding nursing home residents' definitions of quality. *The Gerontologist, 41*(4), 539–545.
Brown, H. (2002). Vulnerability and protection. K202 Course Care Welfare and Community, Workbook Unit 18. Milton Keynes: Open University.
Cattan, M. (Ed.), (2009). *Mental health and well-being in later life*. Maidenhead: Open University Press.
Chatterjee, D. P., Patnaik, P., & Chariar, V. M. (Eds.), (2008). *Discourses on aging and dying*. London: Sage.
Chokkanathan, S. (2009). Resources, stressors and psychological distress among older adults in Chennai, India. *Social Science and Medicine, 68*, 243–250.
Clow, A., & Aitchison, L. (2009). Keeping active. In M. Cattan (Ed.), *Mental health and well-being in later life*. Maidenhead : McGraw-Hill, Open University Press.
Cook, G. A. (2008). Older people actively reconstruct their life in a care home. *International Journal of Older People Nursing, 3*(4), 270–273.

Cornes, M., Peardon, J., & Manthorpe, J. (2008). Wise owls and professors: The role of older researchers in the review of the National Services Framework for older people. *Health Expectation, 11*(4), 409–417.

Datta, A. (2008). Socio-ethical issues in the existing paradigm of care for the older persons: Emerging challenges and possible responses. In Chatterjee et al.

Foucault, M. (1977). *Discipline and punish: The birth of the prison*. London: Allen Lane.

Gadamer, H.-G. (2004). *Truth and method* (3rd ed.). London: Continuum.

Giddens, A. (2009). *Sociology* (6th ed.). Cambridge: Polity press.

Gilleard, C., & Higgs, P. (2005). *Contexts of ageing: Class, cohort and community*. Cambridge: Polity Press.

Gilleard, C., & Higgs, P. (2010). Aging without agency: Theorising the fourth age. *Aging and Mental Health, 14*(2), 121–128.

Grenier, A., & Hanley, J. (2007). Older women and "frailty": Aged, gendered and embodied resistance. *Current Sociology, 55*, 211–228.

Gubrium, J., & Wallace, B. (1990). Who theorises age? *Ageing and Society, 10*(2), 131–150.

Hafford-Letchfield, T. (2010). The age of opportunity? Revisiting assumptions about the life-long learning opportunities of older people using social care services. *British Journal of Social Work, 40*, 496–512.

Harper, M. (2010). The Conservatives: More personalisation and quicker. *Learning Disability Today, 10*(4), 11.

Heidegger, M. (1962). *Being and time: A translation of Sein und Zeit*. NewYork, NY: University of New York Press.

Iliffe, S., Kharicha, K., Kharicha, D., Swift, C., Goodman, C., & Manthorpe, J. (2010). User involvement in the development of a health promotion technology for older people: Findings from the Swish project. *Social Care in the Community, 18*(2), 147–159.

Jamieson, A., & Victor, C. R. (Eds.), (1997). *Researching ageing and later life*. Buckingham: Open University Press.

Jewell, A. (2004). Nourishing the inner being: A spiritual model. In A. Jewell, (Ed.), *Ageing, spirituality and well-being*. London: Jessica Kingsley.

Jewell, A. (Ed.), (2004b). *Ageing, spirituality and well-being*. London: Jessica Kingsley Publishers.

Lamb, F. F., Brady, E. M., & Lohman, C. (2009). Lifelong resiliency learning: A strengths-based synergy for gerontological social work. *Journal of Gerontological Social Work, 52*, 713–728.

Larkin, E., Sadler, S. E., & Mahler, J. (2005). Benefits of volunteering for older adults mentoring at-risk youth. *Journal of Gerontological Social Work, 44*(3), 23–37.

Lawler, S. (2008). *Identity: Sociological perspectives*. Cambridge: Polity Press.

Luppi, E. (2009). Education in old age: An exploratory study. *International Journal of Lifelong Education, 28*(2), 241–276.

Lymbery, M. (2010). A new vision for adult social care? Continuities and change in the care of older people. *Critical Social Policy, 30*(5), 5–26.

Martin, J. (2007). *Safeguarding adults*. Lyme Regis: Russell House Publishing.

Mccabe, L. F. (2006). The cultural and political context of the lives of people with dementia in Kerala, India. *Dementia, 5*, 117–136.

McMunn, A., Nazroo, J., Wahrendorf, M., Breeze, E., & Zaninotto, P. (2009). Participation in socially-productive activities, reciprocity and well-being in later life: baseline results in England. *Ageing and Society, 29*, 765–782.

Moriarty, J., & Butt, J. (2004). Social support and ethnicity. In A. Walker & C. Hennessy (Eds.), *Growing older: Quality of life in old age*. Buckingham: Open University Press.

Morrow-Howell, N., Tang, F., Jeounghee, K., Lee, M., & Sherraden, M. (2005). Maximising the productive engagement of older adults. In M. L. Wykle et al. (Eds.), *Successful aging through the life span*. New York, NY: Springer.

Peace, S. (2002). The role of older people in social research. In A. Jamieson & C. R. Victor (Eds.), *Researching ageing and later life: The practice of social gerontology*. Buckingham: Open University Press.

Phillipson, C. (2007). The "elected" and the "excluded": Sociological perspectives on the experience of place and community in old age. *Ageing and Society, 27*(3), 321–342.

Pickard, S. (2010). The 'Good Carer': Moral practices in late modernity. *Sociology, 44*, 471–486.

Putnam, R. D. (2000). *Bowling alone: The collapse and revival of American community*. Simon and Schuster: New York, NY.

Ray, M. (2007). Redressing the balance? The participation of older people in research. In M. Bernard & T. Scharf (Eds.), *Critical perspectives on ageing societies*. Cambridge: Polity press.

Rorty, R. (1999). *Philosophy and social hope*. London: Penguin.

Rorty, R. (2007). *Philosophy as cultural politics*. Cambridge: Cambridge University Press.

Rowlands, J. (1998). A word of the times, but what does it mean? In H. Afshar (Ed.), *Women and empowerment, illustrations from the third world*. London: MacMillan Press.

Sherratt, Y. (2006). *Continental philosophy of social science: Hermeneutics, genealogy, and critical theory from Greece to the twenty-first century*. Cambridge: Cambridge University Press.

Sibeon, R. (1996). *Contemporary sociology and policy analysis: The new sociology of public policy*. London: Kogan Page/Tudor.

Stones, R. (2005). *Structuration theory*. Palgrave Macmillan: Basingstoke.

Thompson, N. (1998). The ontology of ageing. *The British Journal of Social Work, 28*(5), 695–707.

Thompson, N. (2007). *Power and empowerment*. Lyme Regis: Russell House Publishing.

Thompson, S. (2007). Spirituality and old age. *Illness, Crisis and Loss, 15*(2), 169–181.

Thompson, N. (2010). *Theorizing social work practice*. Basingstoke: Palgrave Macmillan.

Thompson, N. (2011). *Promoting equality: Working with diversity and difference* (3rd ed.). Palgrave Macmillan: Basingstoke.

Townsend, P. (1981). The structured dependency of the elderly: The creation of social policy in the twentieth century. *Ageing and Society, 1*(1), 5–28.

Walker, A., & Hagan Hennessy, C. (Eds.). (2004). *Growing older: Quality of life in old age*. Maidenhead: Open University Press.

Westerhof, G. J., & Tulle, E. (2007). Meanings of ageing and old age: discursive contexts, social attitudes and personal identities. In J. Bond et al. (Ed.), *Ageing* in *society: European perspectives on gerontology*. Sage: London.

Williams, V., Ponting, L., Ford, K., & Rudge, P. (2009). A bit of common ground: Personalisation and the use of shared knowledge in interactions between people with learning disabilities and their personal assistants. *Discourse Studies, 11*(5), 607–624.

Wykle, M. L., Whitehouse, P. J., & Morris, D. L. (2005). *Successful ageing through the lifespan: Intergenerational issues in health*. New York, NY: Springer.

Chapter 8
The Significance of the Findings for Eldercare Practice

Introduction

Given that this research has been underpinned by a concern for social justice (Barry 2005), and that I had presented it to the participants as such, I would have missed an opportunity to promote the integration of theory and practice, and therein the potential for change, if I were not to draw out the significance of the findings for eldercare practice. Furthermore, to do so is to recognise the role that the participants can claim as potential change agents through their role in the production of new knowledge.

While I cannot claim representativeness from such a sample as mine, nevertheless, the study has raised a number of issues that have significance for practitioners organising or delivering care to dependent older people in situations similar to those experienced by the participants. This exploration of reciprocity, and the factors that impede or promote it, has served as a vehicle to bring the spiritual dimension of dependent older people's experience into focus, and supports the existing literature that argues for it to be more prominent on health and social care agendas (Speck et al. 2005; Holloway and Moss 2010). More specifically, the findings have indicated that virtually all of the participants, both in India and the UK, aspired to reciprocity, but felt that this need was not often recognised by those with the power to facilitate it. That the neglect of spiritual well-being was reflected to such an extent reinforces those arguments that it should be researched and theorised more fully, and highlights the need for practitioners to be open to perspectives other than their own 'take' on old age and eldercare, if older people's needs are to be met in meaningful ways.

In terms of the significance of my work for eldercare practice, the findings have highlighted that, if it is to be holistic and life affirming—and dependent older people are to feel that they are recognised as people with strengths as well as weaknesses and as individuals with problems, rather than problems per se—then eldercare practitioners need to have a good understanding of the following concepts and issues relating to them:

- reciprocity;
- spirituality;

- meaning making; and
- dignity.

While recognising their inter-relatedness, I consider each in turn, for clarity of presentation.

Reciprocity

As a term or concept, reciprocity does not appear to figure very prominently in policy documents or training materials relating to the care and support of dependent older people (Bowers et al. 2011). In the sense I have used it,—'usefulness'—it is implied in eldercare initiatives relating to, for example, volunteering, intergenerationality and user involvement—but, as I discussed earlier, these are more often associated with fitter and more independent older people than those who have come to rely on formal care support. As I discuss below, where the focus is on dependent older people's needs and the right to have them met, those needs appear to be more readily conceptualised as physical needs than other aspects. For example, the complaints relating to eldercare in hospitals and residential homes that are chronicled by the organisation *A Dignified Revolution* (http:// www.dignifiedrevolution.org.uk) appear to refer predominantly to assaults on dignity that are related to poor physical care. Similarly, a report published by the Equality and Human Rights Commission into the provision of home care support (Sykes and Groom 2011), while not neglecting personhood, also seems to focus more heavily on physical than spiritual matters, suggesting that the addressing of physical need carries more validity than does spiritual need. In terms of PCS analysis, this may reflect commonly held assumptions that older people do not have those needs, which can help to reinforce the assumption that they are 'less than' other adults.

It could be argued that physical need *should* take priority over other needs as the neglect of physical need may be life-threatening but, to my mind, that is not an argument for neglecting other aspects because, where the need to feel valued through one's usefulness is neglected, then the quality of that life is compromised. Though none of the participants commented that their life had not been worth living because they were not valued as people, there was a good deal of evidence to suggest that their enjoyment of life was significantly diminished by it.

I would support the premise that good quality eldercare is that which, despite pressures on time and finances, and in the context of negative stereotypes of old age abounding, respects the uniqueness of each older person, and also recognises that people and contexts change (Renzenbrink 2004; Thompson 2005; Neuberger 2009). Most of the participants in this study were complimentary about the care they received in that respect, and reflected that practitioners had taken time to get to know them, and their circumstances prior to becoming dependent on them for support. However, this did not extend to exploring their aspirations to remain unique people and valued citizens, and how these could be facilitated in their changed circumstances. This suggests that what reciprocity *meant* to each of them had not been well understood.

Yet, from their narratives, it was significant that being able to reciprocate held different meanings for different participants:

To Megan, reciprocity appeared to mean being a good citizen by utilising what strengths she retained to challenge unfairness:

> they (people) go on the other side, you know. But I won't. I think it's in you, d'you know? It's in your head… call it compassion, I suppose…

To Muthu, reciprocity appeared to mean that he was fulfilling expectations of him as a follower of Hinduism:

> God has created me in that light, that temperament, to help others,

And, for Venkatraman, reciprocity had come to mean looking out for fellow residents:

> Everyone is co-operating… so if I want something, they will come and help me, and if they want something, I will help also… I feel good and proud of me.

And while Eddie understood reciprocity as an obligation to contribute to the well-being of older people beyond those he engaged with in person, Maureen, at the other extreme, appeared to feel no obligation to give something back to others. Yet, this unique significance in people's lives of feeling valued may not necessarily be immediately evident to care practitioners and is likely to take time, and a sensitive relationship built on the respectful 'I-Thou' premise (Buber 1958) to become so. Where that dedicated time and attention is not forthcoming—as was reflected in most of the narratives, and particularly in Ken's, where he reported rushed and instrumental care provision and a lack of respect for his own interpretation of eldercare—it may indicate to the older person that they are not considered worthy of it.

Drawing on PCS analysis, I would argue that this neglect of the importance of reciprocity (especially in the UK) may not necessarily be rooted in the uncaring attitude of any particular practitioner (attitude, behaviour or belief at the P level), but in shared assumptions at the C level that relate to

(a) what is, and is not, considered to be appropriate in terms of old age and expectations of reciprocity; and
(b) the processes whereby these are internalised as how things *should* operate.

And if those shared assumptions incorporate an understanding of dependent old age as decline only, it is understandable that reciprocity is not higher on the eldercare agenda: why would practitioners think it appropriate to devote time to promoting reciprocity in older people's lives if it were not seen to be an appropriate project? But therein lies the significance of the findings for eldercare practice, given the indication that, from the perspective of the older people in this study at least, it *is* considered to be an appropriate project. And where validity is accorded *that* perspective, it provides a challenge to enduring and dominant assumptions of decline and uselessness in old age.

Where a service user perspective is recognised as valid by eldercare practitioners, then coming to understand the meaning making that underpins the unique significance of reciprocity in people's lives has the potential to inform the ways in which it can be facilitated. While appreciating that those providing care for people deemed to be vulnerable are constrained in their practices by having to manage risk while also respecting rights (Stanford 2008; Hothersall 2010), most of the narratives reflected that the opportunities for enhancing their spiritual well-being through engaging in acts of reciprocity could have been easily accommodated in terms of risk management, had the significance of reciprocity to the participants been fully taken on board, and the risks to self-esteem and spiritual well-being recognised as being as much a part of the risk management process as were physical risks.

For example, Eddie's aspiration to help traumatised ex-army colleagues needed only a means of connecting and communicating with them—and the technology to do so is readily and cheaply available.

Similarly, Harold's disappointment at not being able to engage in shared learning opportunities, and the despair he expressed about despotic political regimes, could have been addressed by facilitating adult education opportunities and political activism—again through modern technologies such as e-learning and engagement with online communities, or by enquiring whether another adult learner may have been able to escort him to and from the learning environments he would have like to have been a part of. Muthu had been in a similar situation, requiring only transport to get him from the home he lived into the wider community in which he could have operated as a volunteer in some way, when his health permitted him to do so.

My research did not explore meaning making from the practitioners' perspective, concerning the participants' spiritual need to feel valued, and so I can only suggest that it is likely to have been influenced by, and in turn reinforced, assumptions at the cultural level about what dependent older people should expect and accept as their lot.

I took it as significant that:

(a) those providing care support for the Indian elders tended to take a back seat, in that they allowed the residents to live the life of their choosing, unless it compromised their physical safety. As such, promoting reciprocity did not appear to be high on their eldercare agenda. It may have been significant that this was in the context of a shared Hindu world-view which espoused reciprocity and interdependence as a personal responsibility (Mehta 1997); and
(b) those providing care support for the UK elders did not appear to have promoting opportunities for reciprocity on their agenda either, as evidenced by the participants' comments that their aspirations to it were never explored, thereby making it unlikely that they would be addressed as a response to identified need. For the UK elders, then, the context was different—one where ageism continues to operate at a number of levels to imply that reciprocity is not an 'appropriate' aspiration (Bowers et al. 2011).

That dependent older people attached significance to reciprocity in different ways, and to different degrees, but almost universally across both samples felt spiritually diminished when its importance to them was not recognised by others, is a reminder that:

(a) dependent older people have feelings, and are sensitive to how they are perceived by others; and
(b) furthermore, that their need to feel valued is not diminished by increasing care needs, and may indeed be heightened by assumptions operating at the cultural level (reinforced by and feeding into their personal meaning making in a dialectical fashion) that they constitute a burden.

As the narratives have indicated, all had the capacity, in some way, to counteract that conceptualisation as burden, and to play their part in changing the dynamic of their situation of dependency from one in which they were seen as only capable of receiving care, to one where formal care was seen as a joint venture involving mutual learning.

In this sense, the findings have supported that dependent older people benefit from a 'care-as-relating' dynamic (Bowers et al. 2001) but have extended the analysis further along the psychosocial spectrum to examine how relationships of reciprocity operate in differing cultural settings and within different social structures.

In terms of reciprocity, then, the significance of the findings for eldercare practice lies in their highlighting of the pivotal role that practitioners can play in:

- helping to ensure that reciprocity is recognised as a need (as part of an overall spiritual needs agenda), so that the significance of being valued is not overlooked when interventions to address need are identified;
- facilitating the role that those dependent on formal care can themselves play as agents of change (challenging dominant assumptions at the C level that they should be passive recipients of care), by not impeding opportunities to demonstrate that they can give as well as take;
- ensuring that those older people with whom they work are not denied the social space they need in order to operate, at the very least in some small way, as 'useful' people—even where this may require creativity in matching resource to need;
- reminding colleagues, employing organisations and policy makers of the significance of social time, as well as social space, for the well-being of dependent older people—in the sense that:

    (a) the significance of reciprocity to both parties may change over time: and
    (b) while their life journey may involve elements of decline, this does not preclude aspiration to personal growth.

- not excluding dependent older people from strengths-based therapeutic interventions from which they could potentially benefit. Lamb et al. (2009) have argued that old age should not preclude strengths-based interventions and, in highlighting strengths of which care staff appeared unaware (and how those strengths had made them feel valued in the past), the findings have supported the premise that building on strengths is as appropriate an intervention in this care sector as any other. To suggest otherwise would be to impose one perspective on meaning making over others.

Spirituality

Understanding the significance of reciprocity as an aspect of spirituality relies on an appreciation of spirituality as a broader concept than religiosity, with which it is often equated (Mackinlay 2001; Moss and Thompson 2007; Holloway and Moss 2010). As will have been clear throughout, my own understanding of spirituality is informed by those who equate it more closely with 'human-ness'—the capacity to be aware of who we are, how we see the world and what is important to us (Canda and Furman 1999; Moss 2005). Though not articulated as spirituality either by me in the conducting of the interviews, or by the participants in their narratives, there was evidence that, when understood in these terms, spirituality had been significant for all of the participants.

Given that spiritual questions such as these require us to delve deep into the core of who we consider ourselves to be, it would be tokenistic to treat spirituality as an 'add-on' when assessing need—either informally, as was the case with the Indian elders, or formally in accordance with legal and policy imperatives, as with the UK elders. It is clear that an assessment of what makes people who they consider themselves to be goes beyond what can be ascertained through a one-off tick-box procedure and, even where explored by a sensitive keyworker over a long period of time, may never be fully understood. This is partly because one's spiritual dimension can be hard to articulate, or even to understand oneself.

In terms of my journey through this research, I had anticipated at the beginning that the questions I intended posing to the participants might be difficult for them to answer in an interview situation precisely because I would be asking then to reflect on matters that they may not normally have spent much time thinking about. Because of my background and training, my own self-reflection skills have been honed, but I had been keenly aware that self-reflection may not have been easy for the participants. What I have been able to glean from interviews has no doubt been limited somewhat, both in content and depth because of this, but what I have been able to establish is that spiritual matters *per se* were important because they relate to individual meaning making.

For some, this *had* been related to religiosity, in the sense that living lives in which they could be of service to others had been underpinned by obligations inherent in the Hindu religion they followed. But even in that situation, it can be argued as (Smith 1998) has done, that in distinguishing between Hinduism and Hindu-ness, reciprocity may be understood as an aspect of spirituality that goes beyond religiosity—as a sort of 'feel-good' factor derived from living a lifestyle that is endorsed by tradition, and one's community, as an appropriate way to behave—rather than just an allegiance to religious rules of conduct.

In terms of the significance of the findings for eldercare practice, it may seem that the difference between religiosity and spirituality is of little significance—a matter of terminology and for academic debate only—but I would argue that it is of major significance for a better understanding of the significance of reciprocity in older people's lives if spirituality is understood narrowly as equating only with religious adherence and devoutness. Where spirituality is not understood in the broader sense of incorporating values and aspirations, these aspects of need are unlikely to be explored *precisely* when they are at their strongest. That is, when spiritual well-being is under threat at the point when existential 'certainties' are being challenged by declining health.

For example, Megan had described a constant 'battle' in her life to remain true to herself and her values and, as part of that battle, resisted attempts by others to change her perception of herself as not only self-sufficient, but also someone who could also help others along the way. As such, her narrative reflected that she saw them as threatening what grounded her—reciprocity being one of the constants she aspired to—in a life that was changing because of increasing infirmity. I consider Megan to have been lucky, in that the significance of reciprocity for her spiritual well-being had been recognised, but this was an anomaly in respect of the overall findings which suggested that it was not generally the case.

Of significance for eldercare, then, is the reflecting in the narratives that:

- spirituality is not an 'optional extra' when assessing need—as if only relevant when someone declares membership of a religion which has direct implications for the management of care and death practices—but at the core of eldercare that strives to respond to unique needs in a holistic way;
- spiritual needs may not become evident unless explored in sensitive ways and over time, with the eldercare practitioner acting as guide and support while the older person journeys through a period of self-reflection. This is not to suggest that it requires a specialist role, merely a respect for 'human-ness' and a recognition that those dependent on formal care are not just existing, but continuing to develop as people on a journey through life. I considered that, as a researcher, I was accompanying the participants on a stretch of their spiritual journeying, albeit only on the fringes and for a short time. Eldercare practitioners have the potential to be more enduring companions;
- spirituality incorporates different meaning for different people. Given that the findings have been analysed with regard for meaning making at both an individual level, and in terms of the meaning making that operates at the level of discourses which can be seen to promote a particular viewpoint as 'the truth', it should be clear from the findings that attention needs to be paid to ensuring that assessment of spiritual need reflects older people's interpretation of what gives meaning to their lives, rather than being informed by the practitioner's interpretation of spiritual need. In essence, to understand that spirituality is subjective. More than that, the meanings attached to it are fluid, and therefore assessment of need that is not iterative will never be an adequate response for what is, in effect, a journey not a situation.

Meaning Making

It is clear that the findings have highlighted the need for phenomenological empathy. That is, care provision has the potential to be egocentric, ethnocentric and logocentric if other people's meaning making is not appreciated, or considered of less importance than one's own—if, for example, practitioners understand dependent old age as 'that space where older people are moving towards their death and need only looking after, and on our terms'. As the narratives have indicated, this was not the perception that most of the participants had of their old age, even though they were experiencing infirmity. Though not denying that they were approaching their deaths, for the most part, their narratives were about living (or in some cases not being able to live) fulfilling lives in the meantime.

As we have seen, while ageist discourses operating at the C level continue to associate the care of very dependent older people with decline and 'waiting for death', from a phenomenological perspective, *everyone's* meaning making about life is made in relation to our dying—that is, it is not just older people who live in the face of death. In the context of exploring how the existentialist movement seeks to address the perceived 'meaninglessness' of existence, Tillich (2000) reflects Heidegger's understanding of the implication of knowing that life is finite:

> When Heidegger speaks about the anticipation of one's own death it is not the question of immortality which concerns him but the question of what the anticipation of death means for the human situation. (p. 142)

From an existentialist perspective, then, from the day we are born we are all on that journey towards death. Indeed, life is defined by having parameters of birth and death. It is knowing this, according to Tillich, that requires us to be courageous in facing finitude, and accepting that we need to make life meaningful while we are living it—what he describes as 'the courage to be'. And for many, that meaningfulness will relate to being valued by others. However, the Indian elders, in following the Hindu faith, are likely to have been framing their meaning making in the context of believing that life is *not* finite—in the sense that the soul is considered to be that element which constitutes 'life' and this is presumed not to die when the body does (Chatterjee et al. 2008). It could be argued, therefore, that believing that life is *not* finite can be seen to underpin *their* meaning making in relation to reciprocity, in that having lived in the service of others will have constituted the 'good life' that holds promise for their eventual reward in future incarnations. This aspiration to continue to be of service to their peers or communities, in whatever ways they could manage, was apparent in all but Venkatraman's narratives—and this anomalous finding can perhaps be explained by his having expressly declared that he was withdrawing from worldly interests to focus on self-enlightenment.

The findings have highlighted that it is not enough to recognise that dependent older people have spiritual needs that relate to understanding their place in the world, as do we all, but that these are *unique* spiritual needs. For example, for Venkatraman, the meaning of his old age was articulated as a time for *withdrawing* from the secular world of relationships with fellow human beings, to one of increasing solitude and re-affirming his relationship with his god, through the reading of sacred literature. Megan, on the other hand, articulated her meaning making as relating to *maintaining* contact with her fellow human beings because reciprocity was what had always given her life meaning. What the narratives have highlighted for eldercare, then is that unless practitioners make it their business to find out what makes life meaningful for those they support, they:

(1) run the risk of providing support that is based on assumptions that do not match with the recipients' values and aspirations to reciprocity; and therefore
(2) fail to look creatively at ways it could be facilitated.

Eldercare practitioners are well placed to effect change by being prepared to question established practices that neglect dependent older people's sense of self-worth, and by facilitating the hearing of user perspectives in debates about eldercare policy and practice. However, unless they recognise the impeding of reciprocity as a problem to be addressed, they are unlikely to adopt this role, and unless they listen to the older people they support (which may involve paying attention to what they are not saying, as well as what they are saying), then they may not become aware that it constitutes a problem from another's perspective. The findings have highlighted that this does happen in reality, in that the meaning making significant to the elders in my study was not (except in Megan's case) apparent to those supporting them. And yet, if eldercare is conceptualised as helping elders to live (as far as possible), the life of their own choosing, then eldercare practitioners need to have at least some idea of what that aspired-to life is.

And so, in terms of the significance of the findings for eldercare practice, they can be seen to highlight:

- the need to appreciate that meaning making relating to reciprocity is subjective, but that the significance attached to it can change over time. For example, to Doris, it had been a source of pride that she could be of service to others, but that significance had waned over time. For Megan, Edna and Muthu, however, its significance to them had not, and the loss of opportunity to reciprocate was expressed as being enduring and detrimental to their spiritual well-being.
- that appreciating another's meaning making requires attentive listening (Tehan 2007), which is likely to be time-consuming. It also requires the creating of a safe environment in which meaning making can be explored, and a sensitivity to the fact that what matters to dependent older people may not be easily articulated if they have come to internalise assumptions that (a) they are useless and (b) that promoting opportunities for reciprocity are not the responsibility of eldercare practitioners.
- that distinguishing between eldercare (where dying is implicit) and palliative care in old age (where dying is explicit) may have implications for the impeding or promoting of reciprocity. I applaud the move to extend hospice-type care into community, hospital and residential or nursing home settings (Hall et al. 2011) but wonder whether:
    (a) the hospice philosophy of 'living while dying' may be applied only to those designated as terminally ill, to the detriment of other older people

dependent on formal care for whom reciprocity constitutes part of that living; and
(b) the ongoing spiritual need to feel valued as 'givers' will be recognised within palliative care delivery anyway.

Dignity

It is clear from the findings that being conceptualised only as people who take, rather than people who can give as well as take, caused distress to several of the participants. I would argue that being conceptualised as such constitutes an affront to dignity where dignity is understood as being associated with respect, as is evident in the following definitions: 'The quality of being worthy of self-esteem or honour—worthiness' (Webster New World Dictionary online) and: 'The state or quality of being worthy of honour or respect' (Oxford English Dictionary online).

There is a dignity agenda currently operating in the UK policy-making arena (Baillie and Gallagher 2010) to address shortcomings inherent in eldercare that depersonalises and welfarises older people by conceptualising them as objects of welfare and, in doing so, loses sight of the fact that they are sentient people:

(a) with the capacity to reflect on their circumstances; and
(b) to whom it matters what is made of them by others (Crossley 1996) and whether they are respected for what they can contribute to the world.

If respect is considered to be integral to the dignity agenda, and reciprocity can be seen to generate respect, then it follows that having that respect compromised through no fault of one's own (and only in response to C level assumptions about worth, rather than actual evidence that the respect is undeserved), can therefore be considered to be an affront to dignity which, in itself, constitutes an argument that it should be included on agendas relating to dignity.

In terms of dignity, then, the significance of the findings for eldercare practice relates to the following considerations:

- Recognising that the impeding or promoting of reciprocity in eldercare is pertinent to dignity agendas. It appears that it is increasingly being recognised, in the UK at least, that older people are deserving of dignified care (Neuberger 2009) but the findings suggest that neglecting reciprocity is not associated with compromising dignity. And where it is not conceptualised as such, it will not be addressed as part of that agenda for change.
- Similarly, where the right to feel valued as a human being is not recognised in eldercare practice, the significance of reciprocity in people's lives may not be addressed as part of a rights agenda either. A report published by the Equality and Human Rights Commission and relating to home care services in England (EHRC 2011) suggests that basic human rights are being neglected by service providers and facilitators who, under pressures of time and cost, prioritise the services' needs above those they are providing the services for. It is appreciated

in the report that the right to dignity is one of those that is often compromised. The findings from this study have indicated that, from the participants' perspective, feeling of little or no value *does* constitute an affront to their dignity. Though a small sample, if this were to be replicated in other studies, my argument that the impeding or promoting of reciprocity has a place on both the rights and dignity agendas that are working towards change in eldercare would be reinforced.

- Though there was an emerging dignity agenda evident in the provision of eldercare support in Chennai, this did not appear to relate to reciprocity either, although possibly because it may be informed by assumptions about personal responsibility, as already discussed. What may be more significant is the emergence of care institutions as a response to a downturn in the availability of family support. In cutting people off from existing relationships of reciprocity, and making it difficult to establish new ones, not only can dignity (as it relates to respect) be seen to be compromised, but the association between dependency and old age strengthened in a cultural context which has long espoused interdependency (Mehta 1997).

Conclusion

I have argued that eldercare practitioners are well placed to act as change agents, but only if they recognise that change is needed. If they are working in the same cultural environments in which they were raised, then their understanding of old age is likely to have been influenced by prevailing assumptions about expectations of older people, especially if these are reinforced by organisational cultures which are also underpinned by those assumptions (Linstead 2009). Listening to the perspectives of those older people with whom they work, and noticing that they may differ from their own, evidences critically reflective practice which questions taken-for-granted assumptions and opens up the possibility of change (Thompson and Thompson 2008). The findings have been significant in this respect, not only because of *what* the participants said (though this was very enlightening in terms of highlighting differences in meaning making), but also:

(a) *that* they were saying it (thereby proving that they had something of value to offer when it is assumed by ageist discourses that they do not); and
(b) that they *wanted* to say it—again, contrary to ageist assumptions about dependent old age and decline, they were keen to engage in personal growth and reciprocity through the co-construction of new knowledge.

In doing so, the participants' engagement in this project has reinforced the value of PCS analysis as an explanatory framework by highlighting how change at the P level (participants demonstrating strengths and aspirations which contradict assumptions that dependency precludes reciprocity) has the potential to redefine how dependent old age is conceptualised at the C level—and, because it is a dialectical relationship, their being conceptualised at the C level as having strengths and aspirations makes it more likely that opportunities for reciprocity will be facilitated for individuals as a consequence. That the participants wanted to play a part in that challenge adds further evidence to suggest that dependency in old age may not diminish people's aspirations to remain connected with others and to be valued by them. Apart from a small minority, the narratives reflected a keenness to engage with others, and that being valued by them enhanced their spiritual well-being.

This research has been exploratory and small scale, and so it cannot be assumed that their experiences and meaning making are typical. That is, that depriving dependent older people of the relationships in which reciprocity can flourish, either through omission or commission, necessarily poses a threat to well-being. I would suggest, though, that the findings have provided enough evidence to indicate that meaning making associated with reciprocity needs to be further explored as part of an overall eldercare agenda. Good practice is informed practice (Hopkins 1986; Thompson and Thompson 2008) and what has informed the findings of this study is not the perspective of those removed from the situation, but that of those receiving end of eldercare provision themselves. In the interests of promoting change in eldercare provision, and having promised the participants that I would, I am making their meaning making accessible to eldercare practitioners.

Furthermore, the four-part phenomenologically grounded framework I have presented (Fig. 1.1) has proved to be a useful a tool for testing out my assumptions that social space, social time and meaning making continue to have currency for understanding life as experienced by older people dependent on formal care. Given that it has proved useful for understanding the significance not only of meaning making itself, but of the relationships in which it takes place, I propose it as an analytical tool for integrating theory and practice as they relate to aspects of eldercare other than the focus here on reciprocity. Above all, the findings have highlighted the temporal nature of aspiration to reciprocity. That is, the narratives have reflected that old age is a journey not a destination—part of a life journey in which being able to give as well as take has featured. The concept of journey also features in the work of Holloway and Moss (2010), where they present their 'fellow traveller' model with respect to the role that social workers can play in addressing the spiritual needs of vulnerable people:

> The model presented here is that of the 'fellow traveller', which allows the social worker to go with the service user on their spiritual journey for as long as they feel comfortable and competent to assist. The model is one of travelling alongside, rather than carrying a passenger, although it recognises that the person in need may stumble, lose direction or wish to give up at difficult points in the journey. That said, it is also a reciprocal model in which the traveller may either bring a particular vista to the attention of the other or suggest a direction which strengthens the relationship or proves beneficial to either person. (p. 112)

Having to take from others to a greater degree than before may necessitate changes in how the balance between give and take can be maintained, but the findings have indicated that dependency and reciprocity are not incompatible. The challenge this presents to eldercare practitioners is, to continue the journey analogy, to see themselves as 'personal tour guides'—facilitating opportunities which help restore the give and take balance in ways that are meaningful and not 'other people's things' as so perfectly described by Megan. Where individual meaning making is unrecognised, ignored or trivialised, and eldercare practitioners see themselves as left luggage attendants rather than tour companions, organisers and guides, the depersonalisation which underpins the conceptualisation of institutional care in particular, as 'God's waiting room', will continue to be a powerful and enduring one.

References

Baillie, L., & Gallagher, A. (2010). Evaluation of the Royal College of Nursing's "Dignity: At the heart of everything we do" campaign: Exploring challenges and enablers. *Journal of Research in Nursing, 15*(1), 15–28.
Barry, B. (2005). *Why social justice matters*. Cambridge: Polity Press.
Bowers, B. J., Fibich, B., & Jacobson, N. (2001). Care-as-service, Care-as-relating, care-as-comfort: Understanding nursing home residents' definitions of quality. *The Gerontologist, 41*(4), 539–545.
Bowers, H., Mordey, M., Runnicles, D., Barker, S., Thomas, N., Wilkins, A., et al. (2011). *Not a one-way street: research into older people's experiences of support based on mutuality and reciprocity*. Joseph Rowntree Foundation: York.
Buber, M. (1958). *I and Thou* (2nd ed.). London: Continuum.
Canda, E., & Furman, L. (1999). *Spiritual diversity in social work practice: The heart of helping*. New York: The Free Press.
Cann, P., & Dean, M. (Eds.). (2009). *Unequal ageing: The untold story of exclusion in old age*. Bristol: The Policy Press.
Chatterjee, D. P., Patnaik, P., & Chariar, V. M. (Eds.). (2008). *Discourses on aging and dying*. London: Sage.
Crossley, N. (1996). *Intersubjectivity: The Fabric of social becoming*. London: Sage.
Equality and Human Rights Commission (2011) Close to home: An inquiry into older people and human rights in homecare, EHRC.
Hall, S., Petkova, H., Tsouros, A. D., Constantini, M., & Higginson, I. J. (2011). *Palliative care for older people: Better practices*. Copenhagen, Denmark: World Health Organisation Office for Europe.
Holloway, M., & Moss, B. (2010). *Spirituality and social work*. Palgrave Macmillan: Basingstoke.
Hopkins, J. (1986). *Caseworker*. Pepar: Birmingham.
Hothersall, S. J. (Ed.). (2010). *Need, risk and protection in social work practice*. Exeter: Learning Matters.
Lamb, F. F., Brady, E. M., & Lohman, C. (2009). Lifelong resiliency learning: A strengths-based synergy for gerontological social work. *Journal of Gerontological Social Work, 52*, 713–728.
Linstead, S. (2009). Managing culture. In Linstead et al. (Eds.), *Management and organization: A critical text* (2nd ed.). Palgrave Macmillan: Basingstoke.
Linstead, S., Fulop, L., & Lilley, S. (2009). *Management and organization: A critical text* (2nd ed.). Palgrave Macmillan: Basingstoke.
Mackinlay, E. (2001). *The spiritual dimension of ageing*. London: Jessica Kingsley Publishers.
Mehta, K. (1997). Cultural scripts and the social integration of older people. *Ageing and Society, 17*(3), 253–275.
Moss, B. (2005). *Religion and spirituality*. Lyme Regis: Russell House Publishing.

Moss, B., & Thompson, N. (2007). Spirituality and equality. *Social and Public Policy Review, 1*(1), 1–12.

Neuberger, J. (2009) What does it mean to be old?. In Cann and Dean (Eds.), *Unequal ageing: The untold story of exclusion in old age*. Bristol: The Policy Press.

Renzenbrink, I. (2004). Home is where the heart is. *Illness, Crisis and Loss, 12*(1), 63–74.

Smith, B. K. (1998). Questionning authority: Constructions and deconstructions of hinduism. *International Journal of Hindu Studies, 2*(3), 313–339.

Speck, P., Bennett, K. M., Coleman, P. G., Mills, M., McKiernan, F., Smith, P. T., & Hughes, G. M. (2005) Elderly bereaved spouses: Issues of belief, well-being and support. In Walker (Eds.), *Understanding quality of life in old age*. Maidenhead: Open University Press

Stanford, S. (2008). Taking a stand or playing it safe? Resisting the moral conservatism of risk in social work practice. *European Journal of Social Work, 11*(3), 209–220.

Sykes, W., & Groom, C. (2011). *Older people's experience of home care in England*. London: Equality and Human Rights Commission.

Tehan, M. (2007). The compassionate workplace: Leading with the heart. *Illness, Crisis and Loss, 15*(3), 205–218.

Thompson, S. (2005). *Age discrimination*. Lyme Regis: Russell House Publishing.

Thompson, S., & Thompson, N. (2008). *The critically reflective practitioner*. Palgrave Macmillan: Basingstoke.

Tillich, P. (2000). *The courage to be* (2nd ed.). London: Yale University Press.

Walker, A. (Ed.). (2005). *Understanding quality of life in old age*. Maidenhead: Open University Press.

Chapter 9
Their Journeys and Mine

Introduction

In setting out my rationale for this study I had referred to drawing on individual narratives to build up a bigger picture of the factors that can impede or promote reciprocity in the care of older people dependent on formal care. However, having explored reciprocity from a meaning making perspective, I now consider that, rather than a picture, the findings fit better with the analogy of a sculpture because they have highlighted that, in order to be fully understood, the significance of reciprocity needs to be appreciated from different perspectives.

In this final chapter, I summarise how exploring the lived experience of older people dependent on formal care through the lens of the innovative phenomenologically grounded analytical framework, and my use of PCS analysis in an innovative context have given rise to new insights which add to the existing literature relating not only to reciprocity and eldercare but also to the theorising of old age in general. In doing so, I highlight that new insights have emerged, but also that the study has reinforced the value of working with dependent older people as co-creators of those insights, in a way that:

(a) argues for greater epistemological value to be accorded to service user perspectives; and
(b) highlights engagement in research activity as evidence of the capacity of dependent older people to challenge their conceptualisation as a burden rather than an asset.

Given that this has been an exploratory study, I consider that the findings have addressed the research questions I set myself by highlighting what the participants themselves considered to be problematic. A number of issues emerged from their narratives that provide a challenge to ageist discourses and I move, later in the chapter, to highlight these as foci for future research. And, though I consider the methodological approach and methods employed to have been generally fit for purpose, as a reflective practitioner and researcher, I am aware that by reviewing my research journey, I open myself up to both scrutiny and personal development as a competent researcher. In light of this, and following a review of what the research has not addressed, I offer up my thoughts on how I might have done things differently and better.

I then draw the chapter and book to a close by summarising the benefits at the personal, cultural and structural levels, of challenging the assumption that dependency precludes reciprocity.

The Significance of Social Space, Social Time and Meaning Making

The research questions I set myself relate to the ways in which reciprocity in the lives of older people dependent on formal are either impeded or promoted. The narratives that emerged from the study have suggested that impediments relate largely to ideological assumptions that dependent older people have:

(a) little to give back as a counterpart to what they are given in terms of formal support;
(b) no need of affirming intersubjective feedback, and therefore the connectedness with others that can provide this;
(c) no aspirations to continue to be valued citizens; and
(d) no personhood or feelings.

and that these assumptions underpin the provision of eldercare which:

- reduces opportunities for reciprocity through the processes of ghettoisation and isolation;
- does not recognise the impact of neglecting opportunities to be of value on spiritual well-being; and
- prioritises physical over spiritual well-being as evident in the management of risk.

By drawing on the four-part phenomenological framework to analyse the lived experience reflected in the narratives I have been able to offer food for thought in terms of the extent to which change over time is implicated in the impeding or promoting of reciprocity, as in, for example:

(a) the conceptualisation of old age as different from adulthood—as reflected by the UK elders—or part of it (as reflected by the Indian elders);
(b) the understanding that dependent older people will be able to remember a past in which they had something to give, but also aspire to a future in which they are recognised as valued people;
(c) changes relating to the commodification of care such that, if care is paid for, there may be implications for the obligation to reciprocate; and
(d) changes in the power of discourses, such as those relating to citizenship, medicine and governmentality to inform what is considered to be problematic for society, and how it should be addressed.

In highlighting a mismatch between the meaning making evident in the UK participants' narratives about wanting to reciprocate, and dominant ageist assumptions that older people are 'less than' other adults, and between the Indian elders' narratives about wanting to reciprocate and their being placed in institutions where they are disconnected from their communities, the phenomenological framework has proved to be a useful tool for demonstrating the ways in which reciprocity can be impeded or promoted in relation to meaning making. Furthermore, PCS analysis has been instructive in accounting for *how* that meaning making can be seen to operate and it is to this that I now turn.

A Sensitising Framework

In demonstrating that the meaning making inherent in the narratives bears relation to the cultural and structural contexts in which old age and eldercare are understood in the differing contexts of the UK and India, drawing on PCS analysis has helped to demonstrate the operating of a double dialectic whereby, in the UK context of being isolated from community life and the affirming feedback it offers, not reciprocating at the personal level can be seen to reinforce the commonly held assumption at the C level that it is not appropriate for dependent older people to reciprocate. It has demonstrated too, that this, in turn, can be seen to reinforce the legitimacy of treating those people less favourably at the structural level by marginalising them in terms of income, opportunities for learning and so on. Drawing on PCS analysis has also served to highlight how demonstrating at the P level that dependency need *not* preclude reciprocity has the potential to challenge dominant discourses operating at the C level which, in turn, has the potential to challenge age as a legitimate basis for social division at the S level.

In demonstrating that similar processes appear to be operating in Chennai, whereby;

(a) the personal responsibility for living a valued life being enacted at the P level by the participants can be seen to reflect, and reinforce, the assumption at the C and S levels that old age is part of, not separate from, adulthood; but also that:
(b) change in how old age is conceptualised at the C level (increasingly as burdensome) may be implicated in changes relating to how power and prestige are allocated at the level of social structure, and for the spiritual well-being of dependent older people whose lives are affected as a consequence,

PCS analysis has highlighted that further research into social phenomena such as reciprocity needs to incorporate complexity and dynamism if it is not to be essentialist, atomistic and reductionist. As a sensitising framework it has provided coherence to what might otherwise have amounted to a collection of individual narratives, without losing sight of the significance of those narratives in their own right as windows through which the lived experience of reciprocity can be observed.

> Substantive theories aim to provide us with new empirical information, whereas sensitising theoretical frameworks are intended to furnish general orientations or perspectives; they are intended to equip us with ways of thinking about the world. (Sibeon 1996, p. 4).

In highlighting how the impeding or promoting of reciprocity can be understood with reference to the operating of a double dialectic between agency and culture, and culture and structure, and in conjunction with the four-part phenomenologically grounded framework I have described, the application of PCS analysis has:

(a) moved the understanding of reciprocity in eldercare forward, and further along the psychosocial spectrum than I believe to be evident in the existing literature base;
(b) strengthened the argument that impediments to the promoting of reciprocity are informed by, and therefore have the potential to be addressed by attention to, the extent to which older people are conceptualised at the C and S levels as people with a future yet to be lived—'lives in progress' with a continuing capacity to be 'givers' as well as 'takers'. As I have argued throughout, this potential for a future orientation provides a radical shift away from theorising that is informed by a bio-medical model which tends to neglect that orientation;
(c) provided accessible frameworks as tools to aid the integration of theory and practise in eldercare.

Reflections on the Research Journey

While I consider this to have been a successful project in most respects, nevertheless, there are some things that, on reflection, I might have done differently. For example, while I had spent considerable time and effort trying to ensure that the participants were drawn from settings which reflected typical formal care provision in both the UK and India, I had experienced resistance from care providers in both countries, to allowing me access to interview older people in their care. Though I would argue from a moral perspective, that it was not their decision to make, I could understand their desire to shield people from what they perceived to be potentially intrusive and distressing encounters with a researcher—especially where previously experiences had reinforced those fears. Despite my calling on the help of respected individuals and agencies to argue the case that I would be both entirely open with the potential participants about the nature of the research, and respectful of their decision not to participate if they so chose, the gate-keepers' meaning making contributed towards my samples being drawn from a less diverse range of establishments and care agencies than I would have liked.

In retrospect, then, I have wondered whether spending more time with those whom I called on as 'conduits' to those people whose stories I wanted to hear and analyse, might have proved effective in persuading more of them to allow me to at least present my intended research to those in their care, and allow the older people themselves to decide whether or not to take part. I felt vindicated in my aim to provide a platform for seldom-heard voices in that all of those I approached *did* agree to be involved, and that some expressed gratitude for the opportunity. However, more extensive preparatory work may possibly have allowed more of their voices to be heard.

Recognising that the interview questions called for a degree of introspection and recall, and that I wanted to respect that those frail enough to be deemed in need of formal care might not be able to sustain discussion over a long period, I have felt a little frustrated that some of the issues I hoped to explore had been left unaddressed, or only fleetingly discussed. In retrospect, in my planning of future research partnerships with dependent older people I may consider giving participants more detailed advanced knowledge of the topic than was given in the information sheet accompanying the consent form for this study, should another study require the same degree of introspection.

In light of the discussion above, and had time and opportunity permitted, I might have considered interviewing the participants for a shorter time, but over more than one visit. In the event, the amount of rich data generated met with my expectations and hopes, but I might, for future projects, give more priority to trying to build in more safeguards to ensure that the time available for listening to people's narratives will not be compromised.

Implications of the Findings for Future Research

I have made clear from the outset that this has been an exploratory study and so I had not set out to identify or test out any particular aspect relating to the impeding or promoting of reciprocity. As discussed in Chap. 4 I had included both men and women in the samples in order not to exclude the potential significance of gender. Similarly, I had included those who were paying fees for their care, and those who were not, so as not to exclude class as potentially significant. In the event, the results were inconclusive about the significance of gender or class which, given the small sample size and the broad exploratory range of the study, is not surprising. Nevertheless, the fact that gender and class were implicated in the participants' meaning making, even to a small degree, suggests that more focused research into these aspects could prove fruitful in furthering understanding of why it is that reciprocity and dependency tend not to be seen as compatible.

The following have also emerged from the findings as potential foci for further phenomenologically grounded research:

Diversity within the elderly population: While I made some attempt to incorporate diversity into the samples, this particular study has focused more on differences *between* those experiencing the impeding or promoting of reciprocity in different contexts than on differences *within* those contexts. I would therefore propose that future research into meaning making associated with reciprocity in eldercare might usefully focus more specifically on, for example, gender, ethnicity, world-view, sexual orientation and so on as variables.

The implications of policy change: As has been discussed, it is clear that policy relating to eldercare is in a state of flux—aspects of which (risk management, for example) appear to be having implications for how reciprocity, or its neglect, is being experienced at a personal level. Given that PCS analysis has demonstrated the interconnectedness of personal experience, discourse and social structure, research which focuses more specifically on the relationship between policy and perceived outcomes from a user perspective would, in my opinion, also usefully extend the eldercare knowledge base.

The nature of formal care provision: Although some of the participants were in receipt of domiciliary care and others residential care, it has not been clear from this exploratory study whether the difference can be said to be significant for a better understanding of reciprocity in eldercare, as there were differences both within and across the settings. Given the potential significance for meaning making of organisational culture, a comparative focus on institutional and domiciliary care provision could therefore be enlightening.

Health or social care provision: While all of the participants in this study were being supported by provision from the social care (rather than health) sector, there was some evidence to suggest that their meaning making had been informed, in part, by models which equate old age with sickness. This study did not address health and social care provision as specific variables, and I am not aware of any reciprocity-based studies that have this as a specific focus, especially from a phenomenological perspective.

A matter of perspective: I sought, in this study, to provide a perspective that I understand to have been largely lacking from existing studies of reciprocity in eldercare—that which centres on the meaning that elders dependent on formal care attach to their experience of reciprocity as life-affirming or spiritually diminishing. Implicated in that meaning making was their perception of how practitioners had conceptualised them, but I have not tested out whether those perceptions accurately reflected the practitioners' meaning making. Further research may, therefore, usefully focus on either a practitioner perspective on the factors that impede or promote reciprocity, or a comparison between the two sets of meaning making.

Conclusion

It is my contention that the findings of this study have largely reinforced the findings of existing studies that link reciprocity with physical and mental well-being through its association with social connectedness (Moriarty and Butt, 2004; Davidson et al. 2005; Breheny and Stephens 2009). However, they have also highlighted that, while the implications for spiritual well-being have been shown to be of significance to dependent older people themselves, that particular aspect of well-being has not attracted research interest to any great degree, even though Lustbader's (1991) work in this field set the stage for it some time ago. In highlighting that feeling valued through opportunities to reciprocate *does* hold significance for how older people dependent on formal care make sense of their dependency, and arguing that impediments to that spiritual well-being can be better understood by locating them in a dynamic and sociologically based analytical framework, I consider that I have moved academic debate about the significance of

Conclusion

reciprocity in the lives of older people dependent on formal care forward. In doing so I have reinforced the following premises:

- that the building up of potentially beneficial relationships of mutual support between, rather than within groups social groups ('bridging social capital' as described by Putnam 2000), can be impeded by power relationships operating between dependent older people and care providers and practitioners (described by Fennema and Tillie 2008, as 'vertical ties'), such that it is the practitioners' meaning making about the appropriateness of reciprocity—rather than that of the older people—that tends to inform whether or not opportunities are promoted;
- that reciprocity is contested, in the sense that has been shown to mean different things to different people in different contexts—but that this has not been well understood (Beel-Bates 2007); and
- that the commodification of care thesis (Ungerson 2000; Garey et al. 2002) has validity for understanding how older people make sense of their lives, and therefore for meaning making around the obligation to reciprocate within care relationships.

This research also reinforces, and builds on, research which highlights the significance of past experience for a positive self-image and sense of self-worth (Bornat and Tetley 2011). This, in itself, provides a challenge to the depersonalisation and reification of older people as objects of care, rather than individuals with care needs, but I have built on this by highlighting the significance of aspiration to *retain* that valued self-image and self-worth.

In agreement with Clow and Aitchinson (2009), I contend that providing 'activities' that bear no relation to individual skill, desire or aspiration, does not constitute good quality eldercare practice. In conjunction with neglecting to facilitate what this provided individual lives with meaning, I have argued that this constitutes a lack of respect for personhood and the spirituality it incorporates, and therefore a potential affront to dignity and a source of alienation.

I have no intent to challenge existing studies of reciprocity in relation to eldercare, as those of which I am aware all appear to have a degree of validity, except in terms of the extent to which they tend to neglect a sociological dimension. Rather, I propose that, in locating individual experience within the context of a sensitising theory that integrates, rather than polarises, structure and agency, the personal experiences that this and other studies have reflected can be better understood.

I have highlighted how existing strands of theory do not adequately account for the impeding or promoting of reciprocity in the lives of those dependent on formal care, and offered a combination of the four-part phenomenological framework I have devised, and the insights drawn from PCS analysis, to help address this in a way which:

(1) In the interests of integrating theory and practice, offers accessible analytical frameworks for understanding the sociological and phenomenological significance of not only reciprocity but other phenomena relevant to eldercare;

(2) demonstrates that the positive ageing agenda in gerontological theorising is as relevant to those in situations of dependency on formal care as it is to their more active peers; and
(3) suggests that there is much to be gained by both academe and practice from according personal narratives greater epistemological value.

From my own perspective, the promoting of opportunities for reciprocity in the care of older people dependent on formal care has the potential to produce positive outcomes at three levels:

the personal, because of the implications this has for the maintenance of self-esteem and the ontological security of knowing that one has a valued place-in-the-world;
the cultural, in that, where dependent older people are 'allowed' to operate within broader communities, and not be excluded from former useful roles because of shared assumptions about older people's lack of competence, their demonstrating that they *are* able to give back in those environments (and indeed want to) provides a challenge to those shared assumptions of inferiority; and
the structural, in the sense that society does not cut itself off from a vast resource of experience, knowledge and skill from which it stands to gain by assigning older people to a subordinate position in the social structure.

Using a phenomenological lens to explore the research questions I set myself has brought into sharp focus the extent to which dependent old age has been conceptualised by the participants as a journey—albeit a journey with which they need help to make—rather than as a destination already reached. And, as the focus has been on reciprocity, it seems fitting to end with a reflection on whether the insights of the UK elders may offer food for thought to practitioners in India—who may wish to take on board that they experienced institutional care as placing restraints on their ongoing journey through life, and the spiritual fulfilment to which they felt entitled. Conversely, the insights of the Indian elders may offer food for thought to practitioners in the UK—through their conceptualising of old age as a *part of* life's journey, not *apart from* it.

References

Beel-Bates, C. A., Ingersoll-Dayton, B., & Nelson, E. (2007). Deference as a form of reciprocity among residents in assisted living. *Research on Aging, 29,* 626–643.
Bornat, J., & Tetley, J. (2011). *Oral History and Ageing.* London: CPA.
Breheny, M., & Stevens, C. (2009). "I sort of pay back in my own little way": Managing independence and social connectedness through reciprocity. *Ageing and Society, 29,* 1295–1313.
Castiglione, D., van Deth, J. W., & Wolleb, G. (Eds.), (2008). *The handbook of social capital.* Oxford: Oxford University Press.
Cattan, M. (Ed.), (2009). *Mental health and well-being in later life.* Maidenhead: Open University Press.
Clow, A., & Aitchison, L. (2009). Keeping Active. In Cattan (Ed.), *Mental health and well-being in later life.* Maidenhead: Open University Press.

References

Davidson, K., Warren, L., & Maynard, M. (2005). Social involvement: Aspects of gender and ethnicity. In Walker (Ed.), *Understanding quality of life in old age*. Maidenhead: Open University Press.

Fennema, M., & Tillie, J. (2008). Social capital in multicultural societies. In Castiglione (Eds.), *The handbook of social capital*. Oxford: Oxford University Press.

Garey, A. I., Hansen, K. V., Hertz, R., & Macdonald, C. (2002). Care and kinship: An introduction. *Journal of Family Issues, 23*, 703–715.

Harrington Meyer, M. (2000). *Care work: Gender class and the welfare state*. London: Routledge.

Lustbader, W. (1991). *Counting on kindness: The dilemmas of dependency*. London: The Free Press.

Moriarty, J., & Butt, J. (2004). Social support and ethnicity. In Walker and Hennessy (Eds.), *Growing older: Quality of life in old age*. Maidenhead: Open University Press.

Putnam, R. D. (2000). *Bowling alone: The collapse and revival of American community*. New York, NY: Simon and Schuster.

Sibeon, R. (1996). *Contemporary sociology and policy analysis: The new sociology of public policy*. London: Kogan Page/Tudor.

Ungerson, C. (2000). Cash in care. In Harrington Meyer (Eds.), *Care work: Gender class and the welfare state*. London: Routledge.

Walker, A., & Hagan Hennessy, C. (Eds.), (2004). *Growing older: Quality of life in old age*. Maidenhead: Open University Press.

Walker, A. (Ed.), (2005). *Understanding quality of life in old age*. Maidenhead: Open University Press.

Appendices

Appendix 1

Interview Schedule

Age: *Gender*:
Could you to tell me a little bit about why it is that you have come to need some help from other people to get on with your day to day life. What help do you receive? How long have you been receiving it?

1. I'd like you to think back to the time before you needed this help. What would you say your main role(s) in life has been? What were you skilled at, what did you do that made you feel useful or proud about? What do you know about that could be helpful to others?
2. Now I'd like to bring you back to the present again. Do you think that the people who help with your care now know those things about you?
3. When those people who arranged your care asked about what help you needed did they ask about what you *can* do as well as what you *can't* do?
4. Do those organising and providing your care talk with you about your life, what you'd like to do with it and how they might help with that? Do *you* ever raise such issues with them? If not, why is that?
5. So, although you need help with some things, do you feel that you do, or could do, some of those things you were talking about earlier, in some form at least—so that you're still using the skills you've built up over your life? [Work through appendix and ask those questions]
6. If you haven't been invited to give others the benefit of your knowledge or skills, or it doesn't happen very often, have you any thoughts on why that might be?
7. How does your experience make you feel?

Additional checklist

Your experience or knowledge called on?	Opp offered?	Welcomed?
- organising things—meetings/events/rotas/etc		
- being involved in the training of carers, nurses, social workers, helpers		
- teaching or helping schoolchildren/students young families		
- cooking or advising on cooking or recipes or diet		
- mending/repairing tasks or advising on such tasks.		
- caring for plants or advising on gardening/plantcare		
- helping others through difficult times—providing emotional support/befriending/praying		
- advising on running a household, or doing housework tasks where able to		
- using experience to help others to stay healthy or cope with illness		
- keeping people informed about current affairs/local politics or similar		
- Are there any other areas where you think you could help others because of your knowledge or skills or time you have available?		

Appendix 2

Information Sheet

Thank you for taking the time to consider being part of this research study.

I have a background in both nursing and social work and am now involved in writing about and training social workers. I have always been interested in how people are understood and treated by their fellow human beings and especially in how this can change when people reach what is considered to be old age. I am also concerned about how this might influence those people who are involved in making policies about how people should be helped if they become ill or frail as they become older. Now that I am working at Liverpool Hope University in the United Kingdom, I have the opportunity to undertake a study into how people feel when they become dependent on other people for help in their daily lives. I am keen to hear about people's experiences so that, together, we can use this research to help influence how older people are treated in the future.

If you are willing to let me talk with you for an hour or so I would feel privileged to hear about your experiences as you have become older.

In order to be able to give my full attention to what you have to say I would like to make a tape recording of the conversation so that I could listen to it later. When I have made a written copy of the conversation we have had, that tape will then be wiped clean and, if I refer in my writing to anything you have told me during our conversation, I can promise you that I will not use your real name or give away any details that could be used to identify you.

If you agree to speak with me then I, or the person who gave you this information sheet, will ask you to sign a letter giving your consent. In this way I can be sure that you are happy to take part.

If you do choose to take part I will look forward to meeting you. If not then I thank you for taking the time to read this and wish you well for the future.

Sue Thompson

Appendix 3

CONSENT FORM

Title of Project: Reciprocity Study

Name of Researcher: Sue Thompson

Please tick if you agree

I confirm that I have read the information sheet about this study, or had it read to me, and understand why I have been asked to take part. ☐

I have been given the opportunity to ask questions about what I am being asked to do. ☐

I understand that it is entirely up to me whether I take part in this study and that I am free to withdraw my co-operation at any time without having to explain why. ☐

I am signing below to indicate that I am willing to be interviewed by this researcher, who will have a translator working with her if I do not speak English, or it is not the language I feel most comfortable to speak in. I understand that this is the only place my name will be recorded, that no-one but Sue Thompson will have access to it and that she will only use it to record that I have consented to take part.

Name of participant:
Signature: Date:

Name of Person taking consent
(if different from researcher):
Signature: Date:

Index

A

Affirmation, 3, 15, 52, 54, 57–59, 107, 124, 132, 137, 139, 140, 142, 144, 149, 160, 172
Ageism (ist), 7, 21, 26, 56, 57, 71, 74, 75, 77, 78, 83, 113, 120, 129, 130, 139, 143, 145, 146, 148, 167, 169–171, 192, 200, 204, 207, 211, 212
Alienation, 26, 36, 39, 47, 72, 217
Atomism, 15, 21, 37, 138, 150, 213

B

Biographical continuity/disruption, 45, 46, 71, 161, 168, 189, 201

C

Care-as-relating, 42, 160
Class, 18, 20, 24, 54, 55, 84, 98, 103, 117, 125, 140, 162, 215
Commodification of care, 41, 84, 116, 157, 212, 217
Connectedness, 3, 9, 45, 52, 70, 101, 121, 144, 146, 150, 158, 159, 163, 170, 185, 186, 190, 212, 216
Critical gerontology, 56
Critically reflective practice, 207

D

Dasein, 9, 10
Deference, 49, 116, 155, 159, 163, 164, 193
Deficit models, 3, 5, 12, 179, 189, 190, 192
Dependency, 8, 11, 12, 17, 21, 23, 27, 30, 38, 46, 47, 49, 51–54, 56, 57, 59, 71, 78, 80, 82, 83, 88, 102, 106, 113, 119, 125, 129–132, 139, 142, 144–146, 149, 152, 157, 161, 166, 168–171, 180–186, 188, 201, 207–209, 212, 213, 215, 216, 218
Dialectic of ideas, 22, 138, 148, 180, 213
Dignity, 7, 27, 54, 57, 58, 70, 80, 121, 163, 164, 198, 206–208, 217
Disengagement, 16, 17, 167
Double dialectic, 4, 22, 120, 138, 180–182, 213, 214
Duty/dharma, 17, 41, 51, 56, 102, 111, 118, 120, 121, 127, 128, 140, 141, 158, 165, 168

E

Eldercare, 4, 6, 8, 11, 12, 51, 59, 75, 102, 111, 112, 114, 118, 120–122, 128, 130, 131, 140–142, 147, 149, 156–158, 163, 165, 167, 168, 171, 183–185, 189, 190, 200, 202, 204, 212, 216, 218
Elders, 3, 4, 6–9, 11, 12, 15, 59, 25–27, 46, 49, 60, 71, 80, 81, 84, 92, 97, 98, 106, 108, 111, 114, 119, 122, 124, 129, 132, 137, 143, 144, 149, 158, 161, 165, 167, 169–172, 179, 183, 185–189, 192, 193, 197–203, 205–209, 211–217
Engagement, 19, 27, 43, 53, 56, 57, 88, 123, 139, 147, 153, 163, 200, 207, 211
Existential crisis, 44, 102, 132, 161

F

Formal care/support, 3, 4, 6, 8, 11, 36, 37, 40–42, 47, 50, 52, 55, 58, 71, 73, 85, 92, 98, 108, 139, 143, 150, 154, 157, 171, 185, 189, 193, 201, 208, 215, 217, 218

G

Gender, 18, 21, 24, 54, 55, 84, 107, 108, 126, 140, 143, 215

H

Habitus, 45, 49, 155, 156
Hermeneutic, 16, 18, 19, 29
Hinduism/ness, 17, 35, 53, 141, 165, 199, 202
Horizon (fusion of), 29, 30, 41, 75, 90, 150, 188
Horizontal/vertical ties, 41, 143, 217

I

Institutionalised patterns of power, 11, 23, 36, 54, 57, 58, 72, 97, 102, 128, 165
Intergenerationality, 102, 121, 198
Intersubjectivity, 7, 39, 52, 101

L

Logocentricity, 179
Loss, 7, 47, 55, 86, 119, 152, 205

M

Mastery, 49, 50, 52, 56, 123, 154, 157, 190, 192
Meaningful (ness), 5, 10, 41, 43, 122, 153, 160, 183, 185, 187, 191, 197, 204, 205, 209
Meaning-making, 5, 8–12, 18, 21, 23, 25, 29, 36, 50, 53–55, 57, 59, 71, 73, 75, 77, 97, 101, 138, 145, 172, 183, 185, 198, 201–205, 208, 215–217

P

personal, C—cultural, S—structural (PCS), 4, 5, 12, 21–23, 54, 60, 75, 137, 145, 148, 153, 156, 160, 165, 172, 179–184, 186, 193, 198, 199, 207, 213–215
Personalisation, 27, 171, 190, 191
Personhood, 44, 74, 98, 139, 142, 145, 147, 153, 160, 198, 212, 217
Perspectivism, 5, 19
Phenomenology (ical), 4–6, 8–12, 15, 19, 23, 28, 29, 37, 50, 60, 69, 72, 74, 76, 83, 91, 92, 98, 101, 166, 183–187, 189, 193, 204, 212, 216, 218
Phronesis, 86, 193

Positive ageing, 12, 20, 182, 189, 218
Post-adulthood, 57
Power/empowerment, 4, 6, 11, 19–23, 26, 28, 30, 35, 36, 41, 45, 49, 51, 54, 55, 57, 58, 71, 73, 76–79, 81, 87, 138, 143, 149, 164, 168, 171, 180, 192, 213
Professional gaze, 168
Progressive–regressive method, 10, 152

R

Reciprocity, 3–5, 11, 12, 15, 18, 23, 27, 30, 35–37, 40–44, 48, 50–56, 59, 70–73, 78, 83, 92, 108–114, 121, 128, 130, 138–147, 150, 152, 156, 165, 171, 186, 199, 205, 207, 209, 214, 218
Reflexivity, 70, 73, 74, 78, 79, 108
Resistance, 51, 113, 121, 156, 186, 214
Rights agenda, 206
Risk/risk management, 26, 46, 58, 100, 107, 113, 125–127, 129, 131, 147, 148, 151, 160, 164, 169, 170, 200, 205

S

Self-esteem, 3, 7, 39, 43, 50–53, 55, 71, 90, 102, 107, 110, 114, 124, 139, 143, 144, 152, 156, 160, 162, 190, 206, 218
Sensiti(s)zing theories, 5
Social capital, 3, 37, 40, 41, 46–48, 52, 57, 139, 143, 162, 190, 217
Social model of disability, 56, 169
Social model of vulnerability
Social space, 9, 37–44, 72, 97, 101, 103, 113, 119, 122, 124, 138, 146, 182, 201, 212
Social time, 11, 44, 46, 48, 72, 97, 101, 116, 138, 157, 182, 208, 212
Spirituality, 6, 11, 16, 21, 52, 53, 159, 202, 203, 217

T

Technology, 40, 102, 114, 117, 122, 123, 146, 200
Thrownness, 10, 45, 152, 154, 184
Truth, 16, 19, 29, 74, 166, 192, 203

U

Usefulness, 11, 17, 21, 35, 39, 52, 55, 59, 87, 90, 102, 106, 124, 140, 168, 186, 189, 198

User involvement, 7, 106, 142, 144, 170, 171, 198

V
Virtuous social work research, 80

W
Well-being, 3, 9, 17, 23, 42, 43, 47, 49, 54, 58, 71, 111, 114, 122, 143, 147, 150, 160–165, 170, 181, 200, 203, 212
Worldview, 19, 29, 35, 49, 51, 56, 73, 84, 111, 159, 163, 164, 200, 215